Mathematical Programming and Control Theory

CHAPMAN AND HALL
MATHEMATICS SERIES

Edited by Professor R. Brown
Head of the Department of Pure Mathematics,
University College of North Wales, Bangor
and Dr M. A. H. Dempster,
University Lecturer in Industrial Mathematics
and Fellow of Balliol College, Oxford

Mathematical Programming and Control Theory

B. D. CRAVEN

Reader in Mathematics
University of Melbourne

LONDON

CHAPMAN AND HALL

A Halsted Press Book
John Wiley & Sons, New York

First published 1978
by Chapman and Hall Ltd
11 New Fetter Late, London EC4P 4EE

© 1978 B. D. Craven

Typeset by The Alden Press (London and Northampton) Ltd
Printed in Great Britain at the University Press, Cambridge

ISBN 0 412 15490 0 (cased edition)
ISBN 0 412 15500 1 (paperback edition)

Distributed in the U.S.A. by Halsted Press,
a Division of John Wiley & Sons, Inc., New York

Library of Congress Cataloging in Publication Data

Craven, Bruce Desmond.
Mathematical programming and control theory.

(Chapman and Hall mathematics series)
Includes bibliographical references.
1. Programming (Mathematics) 2. Control theory.
I. Title.
QA402.5.C73   519.4   78-8463
ISBN 0-470-26407-1
ISBN 0-470-26413-6 pbk.

Contents

Preface

In a *mathematical programming* problem, an *optimum* (maximum or minimum) of a function is sought, subject to *constraints* on the values of the variables. In the quarter century since G. B. Dantzig introduced the simplex method for linear programming, many real-world problems have been modelled in mathematical programming terms. Such problems often arise in economic planning – such as scheduling industrial production or transportation – but various other problems, such as the optimal control of an interplanetary rocket, are of similar kind. Often the problems involve nonlinear functions, and so need methods more general than linear programming.

This book presents a unified theory of nonlinear mathematical programming. The same methods and concepts apply equally to 'nonlinear programming' problems with a finite number of variables, and to 'optimal control' problems with e.g. a continuous curve (i.e. infinitely many variables). The underlying ideas of vector space, convex cone, and separating hyperplane are the same, whether the dimension is finite or infinite; and infinite dimension makes very little difference to the proofs. Duality theory – the various nonlinear generalizations of the well-known duality theorem of linear programming – is found relevant also to optimal control, and the

Pontryagin theory for optimal control also illuminates finite
dimensional problems. The theory is simplified, and its
applicability extended, by using the geometric concept of
convex cones, in place of coordinate inequalities.

This book is intended as a textbook for mathematics
students, at senior or graduate level in an American University,
at second or third year level of an honours mathematics
course in England, or at third or fourth year level of the
corresponding course in Australia. The reader requires some
background in linear programming (summarized in Chapters
1 and 3), and a little, very basic, functional analysis (presented
in Chapter 2). Most of the book (except Section 1.10 and
Chapter 5) can, if desired, be read at a finite dimensional level.
Examples and exercises are extensively given in Chapters 2, 4,
5, and 6. However, the book also serves as a monograph on
some recent developments in mathematical programming,
including some not yet in journal articles; and these are
presented at a level accessible to students.

Chapter 1 presents various models of real situations, which
lead to mathematical programming problems. Chapter 2
presents the underlying mathematical techniques; Chapter 3
describes linear programming, as a preparation for the nonlinear theory to follow. The central area is Lagrangean and
duality theory, given in Chapter 4, with various applications.
Chapter 5 extends this to Pontryagin theory and optimal
control problems. Chapter 6 deals with two areas – fractional
programming, and complex programming – which have
figured extensively in recent research, but on which no book
has hitherto appeared. Chapter 7 describes various algorithms
for computing an optimum, and on how these relate to the
theory of earlier chapters; the emphasis is not on the fine
details of computing methods, but rather on the principles
and applicability of the several algorithms. Some specialized
mathematics, which is needed, is given in Appendices.

The expert will note that the use of the 'coordinate-free'
methods with convex cones in place of componentwise
inequalities makes many proofs much shorter, and more
conceptual; various known results become simple particular

cases, not requiring special discussion. The duality theory is partly based on implicit function theorems; these lead directly to certain restrictions without which Lagrangean theorems (F. John, Kuhn–Tucker) are not valid; and give meaning to an optimum, by exhibiting the 'neighbouring' functions with which an optimum function is compared. The Pontryagin theory is presented unconventionally in terms of linear mappings; this enables various extensions – to constraints on trajectories, to partial differential equations – to flow naturally and obviously from the basic Lagrangean theory.

Sections marked * in the left margin may be omitted on first reading. Some literature references are listed at the ends of chapters. These are often journal articles, where the reader may find more details on particular topics.

I am indebted to Mr M. A. Cousland for a substantial improvement to the basic alternative theorem in Section 2.5, and to Miss Ruth Williams for her careful checking of several chapters, and pointing out various amendments. I am also indebted to a referee, Dr Jonathan M. Borwein, for his detailed and helpful comments.

Melbourne, November 1977   B. D. Craven

CHAPTER ONE

Optimization problems; Introduction

1.1 Introduction

If a problem in the real world is described (and thus necess-arily idealized) by a mathematical model, then the problem often calls for maximizing, or minimizing, some function of the variables which describe the problem. For example, it may be required to calculate the conditions of operation of an industrial process which give the maximum output, or quality, or which give the minimum cost. Such calculations differ from the classical 'maximum–minimum' problems of the calculus textbooks, because the variables of the problem are nearly always subject to restrictions – equations or inequalities – called *constraints*. It is therefore not enough to equate a derivative, or gradient, to zero to find a maximum or minimum.

A mathematical problem in which it is required to calcu-late the maximum or minimum (the word *optimum* includes both) of some function, usually of a vector variable, subject to constraints, is called a problem of *mathematical program-ming*. In the past, mathematical programming has often referred only to problems with finitely many variables and constraints; but the same basic theory, as presented here, applies equally to infinite-dimensional problems such as *optimal control*, where the optimum is described by a curve, instead of a finite vector.

This book is concerned with both theory and computational algorithms for mathematical programming, but not with the computer programming which is usually required by anyone who actually wants to calculate an optimum, for a problem of industrial scale. If the functions involved are all linear, the problem is one of *linear programming*. While this book is mainly concerned with *nonlinear* problems, some of the linear theory is described in Chapter 3, since nonlinear methods build on it.

A number of examples of optimization problems are given in the following sections.

1.2 Transportation network

A transportation network may be mathematically modelled, from the viewpoint of operating costs, as follows. Quantities of a product (for simplicity, consider only one product) must be transported from m supply points (e.g. factories) to n destinations (e.g. warehouses, or other factories which use the material). Supply point i can supply a quantity h_i; destination j requires a quantity d_j; the cost of sending x_{ij} units from supply point i to destination j is assumed to be $c_{ij}x_{ij}$, where c_{ij} is constant. The problem is therefore to minimize the total cost

$$\sum_{i=1}^{m} \sum_{j=1}^{n} c_{ij}x_{ij},$$

subject to the constraints $x_{ij} \geqslant 0$,

$$\sum_{i=1}^{m} x_{ij} = d_j \ (j = 1, 2, \ldots, n);$$

and

$$\sum_{j=1}^{n} x_{ij} \leqslant h_i \ (i = 1, 2, \ldots, m).$$

Additional constraints, such as bounds $p_{ij} \leqslant x_{ij} \leqslant q_{ij}$ on the quantity transported between i and j, may be added to the problem. If a particular route i, j is to be excluded altogether, this may be expressed by assigning a very large c_{ij} to that route.

So far, this problem is *linear*, since only linear equations and inequalities occur in it, and the *objective function* $\Sigma \, c_{ij}x_{ij}$ is a linear function of its variables x_{ij}. The problem would become *nonlinear* if *set-up costs* were included; thus if the cost of sending x_{ij} units on route i, j is

$$b_{ij} + c_{ij}x_{ij} \text{ if } x_{ij} > 0, \text{ and } 0 \text{ if } x_{ij} = 0.$$

This effectively introduces integer-valued variables $v_{ij} = 1$ if $x_{ij} > 0$, $v_{ij} = 0$ if $x_{ij} = 0$. (However, *integer programming* methods will not be discussed in this book.)

Observe that the choice of objective function is critical to the result of any optimization (maximization or minimization) problem. Often, the objective function is not clearly given when the problem is first posed, and the applied mathematician, or operations researcher, must formulate an objective function, then check whether it makes sense with the given real-world problem.

1.3 Production allocation model

A factory makes n products on m machines; machine i makes a_{ij} units of product i per hour, at a unit cost of c_{ij}, and has h_i hours available during the planning period. Demands of d_j for product j ($j = 1, 2, \ldots, n$) are to be met, at minimum total cost. The problem is then to minimize

$$\sum_{i=1}^{m} \sum_{j=1}^{n} c_{ij}x_{ij} : x_{ij} \geqslant 0 \ (i = 1, 2, \ldots, m; j = 1, 2, \ldots, n),$$

$$\sum_{i=1}^{m} x_{ij} = d_j \ (j = 1, 2, \ldots, n),$$

$$\sum_{j=1}^{m} a_{ij}^{-1}x_{ij} \leqslant h_i \ (i = 1, \ldots, m).$$

Here x_{ij} units of product j are to be made on machine i; the colon : may be read as *such that*. Of course, additional constraints, such as bounds on some x_{ij}, and perhaps other constraints, relating e.g. to the supply of raw materials, and also set-up costs (since changing a machine from one product to another costs time and money), may be added to the problem. Note that both 1.2 and 1.3 can occur together, as parts of a larger problem of production and distribution; and it may be required to maximize profits instead of minimizing costs.

1.4 Decentralized resource allocation

A large system consists of several smaller subsystems $i = 1, 2, \ldots, k$. Subsystem i is to be operated in a state described by a variable x_i, which must be contained in a set X_i. In order to optimize the system as a whole, a function $\Sigma_{i=1}^{k} f_i(x_i)$ is to be maximized, subject to the constraints $x_i \in X_i$ $(i = 1, 2, \ldots, k)$, and also to overall constraints $\Sigma_{i=1}^{k} q_i(x_i) \leqslant b$ (which might represent, for example, restrictions on the overall supply of raw materials). This is an example of *separable programming*, in which functions of $(x_1, x_2, \ldots, x_k)$ occur only as sums of functions of the separate x_i; this simplifies the computation.

1.5 An inventory model

At time t, a firm holds a stock of its several products, represented by a vector s_t; each component of s_t refers to a different product. Consider only integer times $t = 0, 1, 2, \ldots$. In the time interval $(t, t + 1)$ an amount m_t is manufactured; this manufacture uses, as raw material, some of the existing stock s_t; assume that the amount required is Gm_t, where G is a suitable matrix. (If there were two products, and only the second required some of the first as raw material, then an appropriate matrix G would have form

$$G = \begin{bmatrix} 0 & k \\ 0 & 0 \end{bmatrix}.\Big)$$

During $(t, t + 1)$, a demand requirement of d_t must be met.

A possible cost-minimization model could be:

Minimize $\sum_{t=1}^{n} c^T s_t + a^T m_t$

subject to $s_{t+1} = s_t - Gm_t - d_t$ $(t = 0, 1, \ldots, n-1)$ (*)

and to constraints $p \leqslant s_t \leqslant q$ $(t = 0, 1, \ldots, n-1)$, representing a minimum level ($\geqslant 0$) for the stock, and a maximum level set by available storage. The inequalities apply to each component. The initial stock s_0 is specified.

This is an extremely idealized model. Often both inventory and production, or distribution, would have to be included in the model, to be at all realistic. The inventory model could be extended to allow for the cost of changing the level of production; let $m_t - m_{t-1} = y_t - z_t$, where $y_t \geqslant 0$, $z_t \geqslant 0$, and only one of each pair y_t, z_t is > 0; then an additional cost $k^T y_t + h^T z_t$ could be added to the objective function.

This problem has an analog in continuous (instead of discrete) time, in which (*) becomes $\mathrm{d}/\mathrm{d}t\, s(t) = (I - G)m(t) - d(t)$, and the objective function becomes $\int [c(t)^T s(t) + g(t)^T m(t)]\, \mathrm{d}t$; this allows also the unit costs to vary with time.

1.6 Control of a rocket

At time $t = 0$, a rocket is fired; at time $t > 0$, its position, velocity, and mass of fuel are described by the components of a *state vector* $x(t)$; the programmed rate of burning the fuel is specified by a *control function* $u(t)$. By appropriate choice of $u(t)$ $(0 \leqslant t \leqslant T)$, it is desired to reach the maximum height at time T; or, more generally, some other specified function of $(x(t), u(t))$ $(0 \leqslant t \leqslant T)$, which could be an integral $\int_0^T f(x(t), u(t), t)\, \mathrm{d}t$, is to be maximized, subject to a differential equation of form

$$\frac{\mathrm{d}}{\mathrm{d}t} x(t) = h(x(t), u(t), t)\ (0 \leqslant t \leqslant T),$$

representing Newton's laws of motion applied to the rocket,

and also to some bounds on the control function, such as $|u(t)| \leqslant 1$, $0 \leqslant t \leqslant T$, and to initial conditions on $x(t)$ at $t = 0$.

This is an instance of an *optimal control problem*, in which some function of a *state vector* (or *trajectory*) $x(\cdot)$ and a *control function* $u(\cdot)$, taken over some time interval, is to be minimized or maximized, subject to a differential equation, bounds on the control function, and initial (also perhaps final) conditions. (Some inventory problems in continuous time can be similarly expressed.) Observe that $x(\cdot)$ and $u(\cdot)$ must fall in appropriate classes of functions; thus, in the rocket problem, $x(\cdot)$ must be a differentiable function of time, whereas $u(\cdot)$ can be discontinuous – the rocket motor can be turned on and off – so that $u(\cdot)$ should be a piecewise continuous function.

The planning of a river system, where it is desired to make the best use of the water, can also be modelled as an optimal control problem. Optimal control models are also potentially applicable to economic planning, and to world models of the 'Limits to Growth' kind.

1.7 Mathematical formulation

In each of these problems, it is required to find the minimum or maximum of a function $f(x)$, where x is a vector, or function, in some specified class V, and x is subject to constraints – equations and inequalities – which may be expressed by $x \in K$, where K is the *constraint set*. Thus, in 1.2, x is the vector $\{x_{ij} : i = 1, 2, \ldots, m; j = 1, 2, \ldots, n\}$ and $V = \mathbb{R}^{mn}$, Euclidean space of dimension mn. A similar remark applies to 1.3, 1.4, and 1.5. In 1.6, x is the pair of functions $(x(\cdot), u(\cdot))$; and V is not generally a vector space, but can be made into one by re-formulating the problem. To see this in a simpler case, suppose that V is the space of all real continuous functions x on $[a, b]$, such that $x(a) = 0$ and $x(b) = c \neq 0$. Then the problem can be re-expressed in terms of the continuous functions $y(t) = x(t) - c(t - a)/(b - a)$, which vanish at both endpoints, and hence form a vector space.

Assume therefore that the minimization problem takes the form;

$$\text{Minimize}_{x \in V} f(x) \text{ subject to } x \in K,$$

where the space V is a vector space (of vectors in $\mathbb{R}^k$ for some k, or appropriate functions) – reformulating the problem if necessary.

Maximization or minimization by calculus – equating a gradient to zero – does *not* usually help for a *constrained* problem, since the maximum or minimum is very often on the the *boundary* of the constraint set K. (Note that this is typical for linear programming.) In the simple nonlinear example

$$\text{Minimize}_{x \in \mathbb{R}} f(x) = x^2 \text{ subject to } x \in K = [a, b],$$

where $0 < a < b$, the minimum obviously occurs at $x = a$, where f has gradient $2a \neq 0$. (Thus the constrained minimum is not a *stationary* point of $f(\cdot)$.) The method of Lagrange multipliers, available for minimization subject to equality constraints, can be adapted to inequalities. In this example, $x \in K$ is expressed by the two inequalities $a - x \leqslant 0$, $x - b \leqslant 0$; equating to zero the gradient of the Lagrangean $x^2 + \lambda(a - x) + \mu(x - b)$ gives $2x - \lambda + \mu = 0$; to this equation must be adjoined (see Chapter 4) $\lambda(a - x) = 0$, $\mu(x - b) = 0$, $\lambda \geqslant 0$, $\mu \geqslant 0$; and these have $x = a$ (the minimum), $\lambda = 2a$, $\mu = 0$ as a solution. The Lagrangean for problem 1.4 is similarly

$$\sum_{i=1}^{k} f_i(x_i) + \sum_{i=1}^{k} \lambda_i x_i - \mu \sum_{i=1}^{k} [q_i(x) - b],$$

where λ_i and μ are the Lagrange multipliers. This problem is further discussed in Section 4.3.

The constraint $x \in K$ is usually expressed by a system of equations and inequalities, of which

$$g_i(x) \leqslant 0 \ (i = 1, 2, \ldots, m); \ h_j(x) = 0 \ (j = 1, 2, \ldots, r)$$

is typical; here the g_i and h_j are real functions. This system will be expressed as

$$-g(x) \in S, \ h(x) = 0,$$

where

$$g = \begin{bmatrix} g_1 \\ g_2 \\ \cdot \\ \cdot \\ g_m \end{bmatrix} \text{ and } h = \begin{bmatrix} h_1 \\ h_2 \\ \cdot \\ \cdot \\ h_r \end{bmatrix}$$

are vector functions, and S is the non-negative orthant in $\mathbb{R}^m$. (The minus sign before g is there for later convenience.) In an infinite-dimensional problem, $g_i(x) \leqslant 0$ $(i = 1, 2, \ldots, m)$ may, for example, be replaced by $g(x)(t) \leqslant 0$ for each $t \in [a, b]$; here $g(x)$ is a function of $t \in [a, b]$, and the constraints will be expressed as $-g(x) \in S$, where S is an appropriate subset of a space of functions. The appropriate subsets are *convex cones* (see 1.8 and 2.2), which share the relevant properties of non-negative orthants. (In 1.10, the rocket problem of 1.6 is formulated in these terms.)

Given a problem formulated as

$$\underset{x \in V}{\text{Minimize}} f(x) \text{ subject to } -g(x) \in S, \ h(x) = 0,$$

the questions which arise include the following. Are the constraints consistent, i.e. does there exist some $a \in V$ such that $-g(a) \in S, h(a) = 0$? (A rocket problem may have boundary conditions requiring the rocket to reach the moon, but there may be no such solution.) Assuming consistency, what vectors x are 'near' the vector a (and in which directions from a), and still satisfy the constraints? (Implicit function theorems, and the concept of local solvability, relate to such questions. If a is a minimum, then $f(x) \geqslant f(a)$ for such 'near' x.) What hypotheses are required for conditions of Lagrangean type to be (i) necessary, or (ii) sufficient for a minimum? (Such matters depend on convex sets and functions, and on *linearizing* a nonlinear problem.) How can Lagrangean conditions be represented? (For optimal control problems, this requires representations of dual spaces of some Banach spaces, and leads to ordinary or partial differential equations.) And, as well as theory, what algorithms are available to compute a numerical solution?

These matters are discussed in the following chapters. The main mathematical techniques required are convex sets and functions, theorems on *separating* convex sets by hyperplanes, and the related *alternative theorems*, which state that exactly one of two given systems has a solution. Also needed are Fréchet and other derivatives of functions, and a linearization lemma. Mathematical conventions are listed in 1.8. The underlying theorems are given in Chapter 2; in certain cases, proofs are marked * (omit on first reading), or omitted (when they are standard theorems of functional analysis).

1.8 Symbols and conventions

The usual symbols for logical and set operations will be used, namely $\Rightarrow$ (implies), $\Leftrightarrow$ (if and only if, also written iff), $\forall$ (for all), $\exists$ (there exists); and $\in$ (belongs to), $\cup$ (union), $\cap$ (intersection); \ (set difference; $S\backslash T$ is the set of elements in S but not in T); $\subset$ (inclusion, allowing $=$ as a special case), $\emptyset$ (empty set); $\times$ (cartesian product). Note that $S - T$ does *not* mean set difference.

Vector spaces U, V, W, X, Y, Z are real (except in part of Chapter 6). The spaces considered are $\mathbb{R}^n$ (Euclidean space of n dimensions, with the Euclidean norm $\|\cdot\|$), or various infinite-dimensional normed spaces, such as $C(I)$ (the space of all continuous real functions on the interval I, with the uniform norm $\|x\|_\infty = \sup_{t \in I} |x(t)|$, and $L^p(I)$ (the space of functions whose pth powers are integrable on I, with finite norm $\|x\|_p = [\int_I |x|^p]^{1/p}$). When the space is required to be complete (as are $C(I)$ and $L^2(I)$), a *Banach space* ($=$ complete normed space) will be specified; $\mathbb{R}^n$ is automatically complete. For subsets of a vector space X, and $\alpha \in \mathbb{R}$, $\alpha S = \{\alpha s : s \in S\}$ and $S + T = \{s + t : s \in S, t \in T\}$; note that $S - T = S + (-T) \neq S\backslash T$. Also $\mathbb{R}_+ = [0, \infty) \subset \mathbb{R}$, the real line. A set $S \subset X$ is a *convex cone* if $S + S \subset S$ and $(\forall \alpha \in \mathbb{R}_+) \alpha S \subset S$; (see 2.2).

Given a map (= function) $f: X_0 \to Y$, where $X_0 \subset X$,
$f(T) = \{f(t): t \in T\}$, for $T \subset X_0$, and $f^{-1}(S) =$
$\{x \in X: f(x) \in S\}$, for $S \subset Y$; the symbol $f^{-1}(S)$ does *not*
imply that f is bijective and so possesses an inverse function.
A continuous linear map from X into the real line $\mathbb{R}$ is called
a continuous linear functional on X; geometrically, it
represents a hyperplane (see 2.2). The vector space of all
continuous linear functionals on X is the *dual space* of X, and
denoted X'; the symbol X^* will not be used for dual space,
since it is otherwise needed. Note that $(\mathbb{R}^n)' = \mathbb{R}^n$, but the
distinction of notation between given space and dual space
will be useful. Denote by $L(X, Y)$ the space of all continuous
linear maps from X into Y; in finite dimensions, linear implies
continuous. In expressions such as sM, where $M \in L(X, Y)$
and $s \in Y'$, sM means the composition $s \circ M$. The notation
for maps used here is consistent with matrix-vector notation
if vectors in $X = \mathbb{R}^n$ are written as columns, $M \in L(X, Y)$ is
represented by a matrix (an $m \times n$ matrix if $Y = \mathbb{R}^m$), and
vectors $s \in Y'$ are represented by rows. (Denote the set of all
real $m \times n$ matrices by $\mathbb{R}^{m \times n}$.) The *transpose* of
$A \in L(X, Y)$ is $A^T \in L(Y', X')$ defined by $(A^T y')x = y'Ax$
for all $x \in X$ and all $y' \in Y'$; note that $A^T y' \in X'$ may also
be written $y'A$. If A is represented by a matrix, then A^T is
represented by the transpose matrix.

The *non-negative orthant* in $\mathbb{R}^n$ is the set $\mathbb{R}^n_+ =$
$\{x \in \mathbb{R}^n : x_j \geqslant 0 \ (j = 1, 2, \ldots, n)\}$. Inequalities between
vectors in $\mathbb{R}^n$ are taken component-wise, thus $x \geqslant y$ iff
$(\forall j)x_j \geqslant y_j$ iff $x - y \in \mathbb{R}^n_+$.

A sequence $\{x_n\} \subset X$ *converges* (in norm) to $x \in X$ if
$\{\|x_n - x\|\} \to 0$ as $n \to \infty$; a set $S \subset X$ is *closed* if S contains
the limit of every convergent sequence of its elements; the
interior of $S \subset X$ is the (possibly empty) set, int S, of those
$s \in S$ for which $s + \alpha \mathscr{B} \subset S$ for some $\alpha > 0$, where
$\mathscr{B} \equiv \{x \in X : \|x\| < 1\}$ is the open unit ball; S is *open* iff
$S = \text{int } S$; it is well known that S is open iff $X \backslash S$ is closed. The
closure $\mathscr{C} S$ of S is the intersection of all closed sets containing S; $S = \mathscr{C} S$ iff S is closed; the *boundary* ∂S of S is
$(\mathscr{C} S) \backslash (\text{int } S)$. The *distance* between sets S and T is $d(S, T) =$
$\inf \{\|s - t\| : s \in S, t \in T\}$.

For the dual space X', a different notion of *closed* is useful. A *weak * neighbourhood* of the point $p \in X'$ is any set

$$N(p) = \{y \in X' : |y(x_i) - p(x_i)| < \epsilon \quad (i = 1, 2, \ldots, r)\}$$

specified by finitely many points $x_1, x_2, \ldots, x_r$ in X and a positive number ϵ. (Such neighbourhoods will take the place of balls $\{y : \|y - p\| < \epsilon\}$ in a normed space.) A subset $Q \subset X'$ is *weak * closed* if every point $p \in X' \backslash Q$ has a weak * neighbourhood $N(p)$ which does not meet Q. In consequence, every limit of points in Q must lie in Q, if limit is defined in terms of weak * neighbourhoods. If $X = \mathbb{R}^n$, then $Q \subset X'$ is weak * closed iff closed. Hence, for finite dimensional problems, *weak * closed* may be read simply as *closed*.

Representations will be required for the vectors in certain dual spaces. If $y \in [L^p(I)]'$, with I an interval and $1 < p < \infty$, then there is a function $\hat{y} \in L^q(I)$, where $p^{-1} + q^{-1} = 1$, for which

$$(\forall x \in L^p(I)) \quad y(x) = \int_I \hat{y}(t) x(t) \, \mathrm{d}t.$$

If $p = 1$, then a similar representation applies, with a bounded function $\hat{y}$. These integrals are Lebesgue integrals; however, apart from these cited results, only elementary properties of integrals are used, so that the reader need not be familiar with the Lebesgue theory. All functions occurring will be assumed, without further comment, to be *measurable*, as required by the Lebesgue theory; this is routinely the case in applications.

If $y \in [C(I)]'$, then there is a signed measure μ for which

$$(\forall x \in C(I)) \quad y(x) = \int_I x \, \mathrm{d}\mu.$$

(A signed measure may be regarded as the difference of two measures.) Consequently some (not all) $y \in [C(I)]'$ may be represented by $y(x) = \int_I x(t)\lambda(t) \, \mathrm{d}t$, for a suitable function λ; this fact is used in 1.10.

A function $f : I \to \mathbb{R}^r$, where I is a real interval, is *piecewise continuous* if each component of f is continuous except at finitely many points where there are simple jump discontinuities. A space of such functions may be given the uniform norm $\|f\|_\infty = \sup_{t \in I} |f(t)|$, where $|\cdot|$ here denotes the Euclidean norm in $\mathbb{R}^r$.

Suppose now that $X = X_1 \oplus X_2$, the *direct sum* of two closed subspaces X_1 and X_2. This means that each $x \in X$ is uniquely expressible as $x = x_1 + x_2$ with $x_1 \in X_1$ and $x_2 \in X_2$ (and hence that $X_1 \cap X_2 = \{0\}$), and that the projection which takes x to its corresponding x_1 is a *continuous* linear map. (The latter assumption is nontrivial for Banach spaces.) Then $M \in L(X, Y)$ can be represented by $Mx = M_1 x_1 + M_2 x_2$ where, for $i = 1, 2, M_i$ is the restriction of M to X_i. This equation is conveniently written in partitioned matrix notation as

$$Mx = [M_1 \quad M_2] \begin{bmatrix} x_1 \\ x_2 \end{bmatrix}.$$

If also $Y = Y_1 \oplus Y_2$, then $Mx \in Y$ can be similarly partitioned, giving a further partitioned matrix representation

$$Mx = \begin{bmatrix} y_1 \\ y_2 \end{bmatrix} = \begin{bmatrix} M_{11} & M_{12} \\ M_{21} & M_{22} \end{bmatrix} \begin{bmatrix} x_1 \\ x_2 \end{bmatrix}$$

in terms of suitable continuous linear maps M_{ij}. Note that the multiplication here is by the matrix convention. If X and Y have finite dimensions, then M is represented by a matrix, and the M_{ij} by submatrices; and the multiplication is valid provided that the usual rules of matrix addition and multiplication are followed.

The linear map $M \in L(X, Y)$ has *full rank* if $X = X_1 \oplus X_2$, where $X_2 = M^{-1}(0)$, and M_1 is a bijection of X_1 onto Y. (If X and Y are Banach spaces, not every X_2 has such a corresponding closed subspace X_1. Since M_1 is continuous, the open mapping theorem of functional analysis shows that M_1^{-1} is also continuous.)

The following procedure is required in Chapter 3. Let A be an $n \times n$ matrix. Postmultiplication of A by the partitioned matrix $\begin{bmatrix} I & p \\ 0 & q \end{bmatrix}$ replaces the last column of A by a given

column b if $\begin{bmatrix} p \\ q \end{bmatrix} = A^{-1}b$. Therefore this column replacement

in A premultiplies A^{-1} by $\begin{bmatrix} I & p \\ 0 & q \end{bmatrix}^{-1} = \begin{bmatrix} I & -pq^{-1} \\ 0 & q^{-1} \end{bmatrix}$, a

matrix which differs from the unit matrix only in one
column. (This step is a key one in the simplex method for
linear programming.)

1.9 Differentiability

Let X and Y be Banach spaces, and X_0 an open subset of X.
For present purposes, a map $g : X_0 \to Y$ is *differentiable* if it
can be suitably approximated locally by a linear map. Let
$a \in X_0$. Then $g : X_0 \to Y$ is *Fréchet differentiable* at a if there
is a continuous linear map $g'(a) \in L(X, Y)$ such that, for
$a + x \in X_0$,

$$g(a + x) - g(a) = g'(a)x + \omega(a, x) \qquad (+)$$

where $\| \omega(a, x) \| / \| x \| \to 0$ as $\| x \| \to 0$. It is then well known
that $g'(a)$ is unique, and that the chain rule holds – if also
$h : Y \to Z$ is Fréchet differentiable, and $b = g(a)$, then $h \circ g$ is
Fréchet differentiable, with $(h \circ g)'(a) = h'(b) \circ g'(a)$.

It is convenient to write $\omega(a, x) = o(\| x \|)$ when (as here)
$\| \omega(a, x) \| / \| x \| \to 0$ as $\| x \| \to 0$; then $g(a + x) - g(a) =$
$g'(a)x + o(\| x \|)$. When $X = \mathbb{R}^n$ and $Y = \mathbb{R}^m$, the Fréchet
derivative $g'(a)$ is represented by an $m \times n$ matrix, whose i, j
element is the partial derivative $\partial g_i / \partial x_j$ evaluated at $x = a$.
Many authors use instead the gradient symbol $\nabla g(a)$, and
mean by it sometimes $g'(a)$, and sometimes $g'(a)^T$. Existence
of $g'(a)$ implies existence of each $\partial g_i / \partial x_j$, but not conversely;
if all the $\partial g_i / \partial x_j$ are continuous in x, then the Fréchet
derivative exists.

The map $g : X_0 \to Y$ is *continuously Fréchet differentiable*
on X_0 if g is Fréchet differentiable at each $x \in X_0$, and $g'(x)$

is a continuous function of x. It then follows (see Appendix A.1) that, for each $\epsilon > 0$, there exists $\delta(\epsilon) > 0$, such that

$$g(x + y) - g(x) = g'(a)y + \xi(x, y)$$

where $\|\xi(x, y)\| < \epsilon \|y\|$ whenever $\|x - a\| < \delta(\epsilon)$ and $\|y\| < \delta(\epsilon)$. In this case, the linear map $g'(a)$ leads to a local approximation to $g(\cdot)$ which is uniform over a neighbourhood of a.

The map g is *linearly Gâteaux differentiable* at a if there is $g'(a) \in L(X, Y)$ such that, for each $x \in X$,

$$\alpha^{-1}[g(a + \alpha x) - g(a) - \alpha g'(a)x] \to 0 \ \text{ as } \ \alpha \downarrow 0 \text{ in } \mathbb{R}_+,$$

(or as $\alpha \to 0$). Note that Fréchet differentiability implies linear Gâteaux differentiability.

*1.10 Abstract version of an optimal control problem

An optimal control problem, such as in 1.6, first paragraph, can be expressed in the abstract form of 1.7; and then Lagrange-multiplier theory can be applied, to describe the optimum. The following formal account will be validated in Chapter 5, and the form of the solution discussed.

Consider an optimal control problem, to minimize

$$F(x, u) = \int_0^T f(x(t), u(t), t) \, \mathrm{d}t$$

subject to the differential equation

$$\frac{\mathrm{d}}{\mathrm{d}t} x(t) = m(x(t), u(t), t) \ (0 \leqslant t \leqslant T),$$

and the constraints

$$g(u(t), t) \in S \ \text{ and } \ n(x(t), t) \in V \ (0 \leqslant t \leqslant T).$$

(The last constraint is absent in 1.6, but arises e.g. in inventory problems.) Assume here that $x(t) \in \mathbb{R}^n$, $u(t) \in \mathbb{R}^k$; f, m, g, n are continuously differentiable functions; and that

---

*   This section may be read at any stage between now and Chapter 5.

boundary conditions $x(0) = x(T) = 0$ are imposed. Let $S \subset \mathbb{R}^r$ and $V \subset \mathbb{R}^h$ be convex cones. (Note that a constraint $(\forall t \in [0, T]) \, \|u(t)\| \leqslant 1$, where $\| \cdot \|$ means here the Euclidean norm in $\mathbb{R}^r$, can be expressed $g(u(t)) \equiv 1 - \|u(t)\|^2 \in \mathbb{R}_+$.)

Appropriate vector spaces of functions must be specified. Let $u \in U$, the space of piecewise continuous functions (see 1.8) from $I = [0, T]$ into $\mathbb{R}^k$, with the uniform norm $\| \cdot \|_\infty$. Denote by W the space of piecewise continuous functions from I into $\mathbb{R}^n$, with the uniform norm. Let $x \in X$, the space of continuous functions from I into $\mathbb{R}^n$, such that each $x \in X$ is the integral of a function $w \in W$, and $x(0) = x(T) = 0$. Then $x(t) = \int_0^t w(s) \, \mathrm{d}s$, which may be expressed as $w = Dx$, where $D = \mathrm{d}/\mathrm{d}t$ except at discontinuities of w; the linear map $D : X \to W$ is made continuous, by giving X the norm $\|x\| = \|x\|_\infty + \|Dx\|_\infty$.

The differential equation for $x(t)$, with initial condition, expressed as

$$x(t) = \int_0^t m(x(s), u(s), s) \, \mathrm{d}s \quad (x \in X, u \in U)$$

may then be written as $Dx = M(x, u)$, where the map $M : X \times U \to W$ is defined by $M(x, u)(t) = m(x(t), u(t), t)$. (The differential equation has now been slightly extended, to allow a finite number of exceptional points where x is not differentiable.)

Let Q and P denote the spaces of piecewise continuous functions from I into $\mathbb{R}^r$ resp. $\mathbb{R}^h$: define the convex cones $K \subset Q$ and $J \subset P$ as

$$K = \{q \in Q : (\forall t \in I) q(t) \in S\};$$
$$J = \{p \in P : (\forall t \in I) p(t) \in V\}.$$

Define the maps $G : U \to Q$ and $N : X \to P$ by $G(u)(t) = g(u(t), t)$ and $N(x)(t) = n(x(t), t)$ $(t \in I)$. Then the constraints on u and x are expressed as $G(u) \in K, N(x) \in J$.

The given minimization problem is now expressible as

(OC): $\underset{x \in X, u \in U}{\text{Minimize}} \{F(x, u) : Dx = M(x, u), G(u) \in K, N(x) \in J\}.$

Note that U is not a Banach space; it may be enlarged to a Banach space $\tilde{U}$, the *completion* of U, by adjoining suitable limits of sequences; and similarly for the other spaces. (Note also that the problem could be reformulated with the uniform norms replaced by $L^1(I)$ or $L^2(I)$ norms.)

Since m is continuously Fréchet differentiable,

$$m(x(t) + z(t), u(t), t) - m(x(t), u(t), t)$$
$$= m_x(x(t), u(t), t)z(t) + \zeta(t)$$

if $x, x + z \in X$, where m_x denotes partial derivative, and $\|\zeta(t)\| \leqslant \epsilon \|z(t)\| \leqslant \epsilon \|z\|_\infty$ if $\|z\|_\infty < \delta(\epsilon)$. Consequently M is partially Fréchet differentiable with respect to x, with derivative $M_x(x, u)$ given by $M_x(x, u)z(t) = m_x(x(t), u(t), t)z(t)$. Also $F(x, u)$ is Fréchet differentiable with respect to x, with derivative $F_x(x, u)$ given by $F_x(x, u)z = \int_0^T f_x(x(t), u(t), t)z(t) \, dt$; here

$$\|F(x + z, u) - F(x, u) - F_x(x, u)z\| \leqslant \epsilon \|z\|_\infty \text{ if } \|z\|_\infty < \delta(\epsilon/T).$$

And similarly for the other functions.

For the problem (OC), define a *Lagrangean*

$$L(x, u; \tau, \bar{\lambda}, \bar{\mu}, \bar{\nu})$$
$$= \tau F(x, u) - \bar{\lambda}(Dx - M(x, u)) - \bar{\mu}G(u) - \bar{\nu}N(x),$$

where $\tau \in \mathbb{R}_+, \bar{\lambda} \in W', \bar{\mu} \in Q', \bar{\nu} \in P'$ are *Lagrange multipliers*. Often τ can be taken as 1, but not always. Later theory, in Section 4.4, shows that, under suitable hypothesis, necessary (and sometimes also sufficient) conditions for (OC) to attain a minimum at $(x, u) = (\xi, \eta)$ are that the Fréchet derivative of L is zero at (ξ, η), together with transversality conditions (discussed later). The zero Fréchet derivative gives:

$$\tau F_x - \bar{\lambda}(D - M_x) - \bar{\nu}N_x = 0;$$
$$\tau F_u - \bar{\lambda}(-M_u) + \bar{\mu}(-G_u) = 0,$$

(#)

where the partial derivatives F_x etc. are evaluated at (ξ, η).

The Lagrange multipliers are not all zero, and satisfy $\bar{\mu} \in K^*$, $\bar{\nu} \in J^*$.

Now suppose that $\bar{\lambda}$ can be represented (see 1.8) by a function $\lambda(\cdot)$, where

$$(\forall z \in X) \ \bar{\lambda}(Dz) = \int_0^T \lambda(z)Dz(t) \, dt.$$

(This involves some restriction on $\bar{\lambda}$, to be validated later.) Then, integrating by parts, and using the boundary conditions $z(0) = z(T) = 0$,

$$Dz = -\int_0^T [D\lambda(t)]z(t) \, dt.$$

The first equation of (#) then shows that, for each $z \in X$,

$$\int_0^T \{\tau f_x(\xi, \eta, t) + \lambda(t)m_x(\xi, \eta, t) + \lambda'(t)$$

$$- \nu(t)n_x(\xi, t)\}z(t) \, dt = 0,$$

where $\bar{\nu}$ is represented by the function $\nu(\cdot)$, and the Fréchet derivatives F_x etc. are represented, as above, by integrals. Therefore $\{\ldots\} = 0$. The second equation of (#) may be similarly treated, assuming that μ is represented by a function $\bar{\mu}(\cdot)$.

A minimum of (OC) at (ξ, η) therefore implies two differential equations; the transversality conditions, which are

$$\bar{\mu}G(\eta) = 0, \quad \bar{\nu}N(\xi) = 0,$$

lead to two more equations. Hence a minimum of (OC) implies the following system:

$$(OCDE) \begin{cases} \tau f_x(\xi, \eta, t) + \lambda(t)m_x(\xi, \eta, t) + \lambda'(t) \\ \quad - \nu(t)n_x(\xi, t) = 0; \\ \tau f_u(\xi, \eta, t) + \lambda(t)m_u(\xi, \eta, t) - \mu(t)g_u(\eta, t) = 0; \\ \mu(t)g(\eta, t) = 0; \ \nu(t)n(\xi, t) = 0; \ (0 \leqslant t \leqslant T). \end{cases}$$

Since the equations (OCDE) are solvable for $\lambda(\cdot), \mu(\cdot), \nu(\cdot)$ as piecewise continuous functions, the representations assumed for $\bar{\lambda}, \bar{\mu}, \bar{\nu}$, are now validated.

Since $\bar{\mu} \in K^*$, $\int_0^T \mu(t)z(t)\,dt \geqslant 0$ whenever $(\forall\, t \in [0, T])$ $z(t) \in S$. If $S = \mathbb{R}_+^r$, then $(\forall\, t)\mu(t) \geqslant 0$ is the requirement on μ. Similar comments apply to ν.

References

Craven, B.D., and Mond, B. (1973), Transposition theorems for cone-convex functions, *SIAM J. Appl. Math.*, **24**, 603–612.

Dieudonné, J. (1960), *Foundations of Modern Analysis*, Academic Press, New York.

Gray, P., and Cullinan-Jones, C. (1976), Applied Optimization — a Survey, *Interfaces*, **6**(3), 24–46.

Simmons, G.F. (1963), *Introduction to Topology and Modern Anal Analysis*, McGraw-Hill/Kogakusha, Tokyo.

Notes

Gray and Cullinan-Jones includes a list of published applications of non-linear programming to real-world problems. For background in functional analysis, as required in 1.8 and 1.10, see Simmons; for Fréchet derivatives (1.9), see Chapter 8 of Dieudonné. Section 1.10 extends some results in Craven and Mond (1973).

CHAPTER TWO

Mathematical techniques

2.1 Convex geometry

Let X be a (real) vector space (e.g. $\mathbb{R}^n$ or $C(I)$). A linear combination $\lambda_1 x_1 + \lambda_2 x_2 + \ldots + \lambda_r x_r$ of the vectors $x_1, x_2, \ldots, x_r$ in X is called a *convex combination* of these vectors if the real numbers λ_i satisfy $(\forall i)\lambda_i \geqslant 0$, and $\Sigma_{i=1}^r \lambda_i \equiv 1$. The *convex hull* co E of a set $E \subset X$ is the set of all convex combinations of finite sets of points in E. In particular, co $\{x, y\}$ is the straight-line segment $[x, y]$ joining the points x and y. A set $E \subset X$ is *convex* if $[x, y] \subset E$ whenever $x, y \in E$. (Equivalently, by an induction on the number of points, E is convex iff $E = $ co E.)

† *Exercise.* If E and F are convex sets, show that $E \cap F$ and $E + F$ are also convex; but $E \cup F$ is *not* usually convex (consider two line-segments in the plane.) Show that $\mathbb{R}_+^n$ is convex. If $A \subset G \subset X$ with G convex, show that $A \subset $ co $A \subset G$, and hence that co A is the intersection of all convex sets containing A

Assume now that X is normed. The closure $\overline{\text{co}}E$ of co E is called the *closed convex hull* of $E \subset X$.

---

† The exercises must not be omitted. Various necessary ideas are introduced only in exercises.

If $E \subset X$ is convex, a point $p \in E$ is an *extreme point* of E if

$$p \in [x, y] \subset E \Rightarrow p = x \ \text{or} \ p = y.$$

Exercise. Show that the square (area) $E \subset \mathbb{R}^2$ whose vertices are $\begin{bmatrix} 1 \\ 1 \end{bmatrix}, \begin{bmatrix} 1 \\ -1 \end{bmatrix}, \begin{bmatrix} -1 \\ 1 \end{bmatrix}, \begin{bmatrix} -1 \\ -1 \end{bmatrix}$ has these four points as its (only) extreme points; and that $E = \text{co extr} E$, where $\text{extr} E$ denotes the set of extreme points of E.

The line through the points $x, y \in X$ is $\{\lambda x + (1 - \lambda)y : \lambda \in \mathbb{R} \}$. A set $E \subset X$ is a *linear variety* (or, more precisely, *affine variety*) if it contains the line through each pair of its points. Let E be a linear variety, and let $z \in E$; then $M = E - \{z\}$ is a subspace of X; for let $x, y \in M$ and $\alpha \in \mathbb{R}$, then $x = \tilde{x} - z, y = \tilde{y} - z$, with $\tilde{x}, \tilde{y} \in E$, and so $\alpha x = [\alpha \tilde{x} + (1 - \alpha)z] - z \in M$ and $\frac{1}{2}(x + y) = [\frac{1}{2}\tilde{x} + (1 - \frac{1}{2})\tilde{y}] - z \in M$. Conversely, if M is a subspace, and z a fixed vector, it is readily shown that $M + \{z\}$ is a linear variety. The *affine hull* aff E of a set $E \subset X$ is the intersection of all linear varieties containing E; since any intersection of linear varieties is a linear variety, aff E is a linear variety.

Exercise. Show that aff E is the set of all vectors $\Sigma_i \lambda_i x_i$ where each $x_i \in E$, each $\lambda_i \in \mathbb{R}$, $\Sigma_i \lambda_i = 1$. (Here Σ denotes finite sum.)

Exercise. If f is a nonzero continuous linear functional on X (thus $0 \neq f \in X'$) show that $H = f^{-1}(0)$ is a closed subspace of X.

Since $0 \neq f, f(b) \neq 0$ for some $b \in X$; then $x - [f(x)/f(b)]b \in H$, for each $x \in X$, so $X = H + \{\lambda b : \lambda \in \mathbb{R} \}$, so the only subspace properly containing H is X. Such a subspace H is called a *hyperplane*. A set $\{x \in X : f(x) \geq \beta\}$, for some $\beta \in \mathbb{R}$, is called a *closed halfspace* of X.

Exercise. Show that $f \in (\mathbb{R}^n)'$ iff $(\forall x \in \mathbb{R}^n)f(x) = (b, x)$ for some $b \in \mathbb{R}^n$, where the inner product $(b, x) = b^T x$ in column-vector terms. Hence each closed halfspace has the form $\{x \in \mathbb{R}^n : b^T x \geqslant \beta\}$.

2.1.1 Lemma. Given $A \in \mathbb{R}^{m \times r}$ and $b \in \mathbb{R}^m$, let $Q = \{x \in \mathbb{R}_+^r : Ax = b\} \neq \emptyset$; for $x \in Q$, denote A_x the matrix of those columns of A whose corresponding component of x is > 0. Then A_x has linearly independent columns (denote this by $A_x \in \mathrm{LIC}$) iff $x \in \mathrm{extr}\, Q$.

Proof. If $x \in Q$ and $A_x \notin \mathrm{LIC}$, then $A_x u = 0$ for some $u \neq 0$; set $h = \begin{bmatrix} u \\ 0 \end{bmatrix}$; then $y = x + \epsilon h \geqslant 0$ and $z = x - \epsilon h \geqslant 0$ (componentwise) for some $\epsilon > 0$; since $x = \frac{1}{2}(y + z)$ and $Ay = Az = Ax = b$, $x \notin \mathrm{extr}\, Q$. If $A_x \in \mathrm{LIC}$, and $x = \lambda y + (1 - \lambda)z$ with $y, z \in Q$ and $0 \leqslant \lambda \leqslant 1$, then the components of y and z *not* corresponding to columns of A_x are zero (they are $\geqslant 0$, and those of x are zero); and the remaining components are the same for y and z (since $A_x \in \mathrm{LIC}$, $A_x u = b$ determines u uniquely); hence $y = z = x$, so that $x \in \mathrm{extr}\, Q$.

2.1.2 Theorem. Let B be the intersection of finitely many closed half-spaces in $\mathbb{R}^n$, and let B be bounded; then $B = \mathrm{co\, extr}\, B$.

* *Proof.* Since B is bounded, the origin may be shifted so that $B \subset \mathbb{R}_+^n$; then $B = \{v \in \mathbb{R}_+^n : Cv \geqslant b\}$ for some matrix C and vector b. Map B affinely one-to-one onto $Q = \{(v, w) \subset \mathbb{R}_+^n \times \mathbb{R}_+^m : Cv - w = b\}$. Then $(v, w) \in \mathrm{extr}\, Q \Rightarrow x \in \mathrm{extr}\, B$, and the Lemma shows, with $A = [C \ -I]$ and $r = m + n$, that $A_x \in \mathrm{LIC}$ iff $x \in \mathrm{extr}\, Q$. If $x \in Q$ but $x \notin \mathrm{extr}\, Q$, then $A_x \notin \mathrm{LIC}$, so $Ah = 0$ for some $h \neq 0$. If $h \geqslant 0$ then, for each $\alpha \in \mathbb{R}_+$, $A(x + \alpha h) = Ax = b$, so $x + \alpha h \in Q$, contradicting the boundedness of Q (from that of B), so some component of h is < 0; similarly some component of h is > 0. So $\rho_1 = \min\{x_i/h_i : h_i > 0\} > 0$ and $\rho_2 = \min\{-x_i/h_i : h_i < 0\} > 0$. Hence $y = x - \rho_1 h \geqslant 0$, $z = x + \rho_2 h \geqslant 0$; $y, z \in Q$; each of

y, z has (at least) one less positive component than x; and $x = \lambda y + (1 - \lambda)z$ where $\lambda = \rho_1/(\rho_1 + \rho_2)$. If y, or z, $\notin \text{extr}\, Q$, this process can be applied to y, or z. In at most 2^r such steps (since the dimension is r), extreme points of Q (and so of B) are reached, noting that if a vector u with only one positive component is reached, then $A_u \notin \text{LIC}$ trivially. Combining these results, $x \in Q$ is a convex combination of the finite set of ($\leqslant \binom{r}{m}$) extreme points which have been found.

2.1.3 Remark. Let $f \in (\mathbb{R}^n)'$; since B (as in the theorem) is bounded closed, the (continuous) f attains a minimum on B, say at $x = p \in B$. Now $p = \Sigma_i \lambda_i e_i$, where the e_i are extreme points of B, and $\lambda_i \geqslant 0$, $\Sigma_i \lambda_i = 1$. If $(\forall i) f(p) < f(e_i)$ then a contradiction results, since $f(p) = \Sigma \lambda_i f(e_i) > (\Sigma \lambda_i) f(p) = f(p)$. So the minimum is attained at some extreme point (and perhaps also at some non-extreme points).

2.2 Convex cones and separation theorems

A set $S \subset X$ is a *convex cone* if $(\forall \lambda \in \mathbb{R}_+)\lambda S \subset S$, and $S + S \subset S$.

Exercise. Show that $\mathbb{R}^n_+$ is a convex cone (this example motivates the definition).

2.2.1 Remark. A convex cone is a convex set. If S is a convex cone, then the relation $\geqslant$ defined by $x \geqslant y \Leftrightarrow x - y \in S$ is a *pre-order* (meaning $x \geqslant x$, and $(x \geqslant y$ and $y \geqslant z) \Rightarrow x \geqslant z$), and also satisfies $x \geqslant y \Rightarrow x + z \geqslant y + z$ for each z. The pre-order becomes a *partial order* (requiring also $(x \geqslant y$ and $y \geqslant x) \Rightarrow x = y$) if S is *pointed* ($S \cap (-S) = \{0\}$). If $S = \mathbb{R}^n_+$ then $x \geqslant 0$ iff each component of x is $\geqslant 0$.

The *dual cone* (or *polar cone*) of S is the convex cone

$$S^* = \{y \in X' : (\forall s \in S)\, y(s) \geqslant 0\}.$$

If $S \subset \mathbb{R}^n$ then $S^* \subset (\mathbb{R}^n)' = \mathbb{R}^n$. The relation of convex cone to dual cone is *not* the same as that of vector space to dual space, hence the different notations S^*, X'. Some authors replace $\geqslant 0$ by $\leqslant 0$, thus reversing signs in many equations; some use $\geqslant$ and $\geq$ with different meanings – but not this book.

Exercise. The sector in $\mathbb{R}^2$ specified by $x_1 \geqslant 0$ and $|x_2| \leqslant c|x_1|$ is a convex cone, pointed if the constant $c < \infty$; the cone has nonempty interior (the cone is then called *solid*) if $c > 0$. Show that the dual cone is also a sector.

Exercise. If $S \subset \mathbb{R}^n$ is a convex cone, show that S^* is a *closed* convex cone (closed meaning a closed subset of $\mathbb{R}^n$).

2.2.2 Lemma. If S and T are convex cones, then $S^* \cap T^* = (S + T)^*$.

Proof. If $u \in S^* \cap T^*$ then $u(S) \subset \mathbb{R}_+$ and $u(T) \subset \mathbb{R}_+$, so $u(S + T) \subset \mathbb{R}_+$; hence $S^* \cap T^* \subset (S + T)^*$. Conversely, $0 \in S \Rightarrow T = \{0\} + T \subset S + T \Rightarrow (S + T)^* \subset T^*$; similarly $(S + T)^* \subset S^*$.

The following *separation theorem* for convex sets is a version of the Hahn–Banach theorem in functional analysis (see References). It is fundamental to many theorems on mathematical programming.

2.2.3 Separation theorem. Let K and M be convex subsets of a normed space X, with $K \cap M = \emptyset$. If K is open, then there exists nonzero $g \in X'$ such that $\sup_{x \in M} g(x) \leqslant \inf_{x \in K} g(x)$. If K is closed, and M consists of a single point b, then there exists nonzero $g \in X'$ such that $g(b) < \inf_{x \in K} g(x)$; the latter result also holds with the roles of X and X' interchanged, and $K \subset X'$ assumed weak * closed.

Thus the linear variety $V = g^{-1}(\alpha)$, where $\alpha = \inf_{x \in K} g(x)$, *separates* K (on one side of V) from b (on the other side), when $M = \{b\}$. Note also (see Appendix A.2) that S^* is weak * closed in X' for any convex cone $S \subset X$, whether or not S is closed. In finite dimensions, a set is weak * closed iff closed.

2.2.4 Cone corollary. Let $S \subset X$ be a closed convex cone; let $b \in X \backslash S$. Then there exists $g \in S^*$ such that $g(b) < 0$. (Consequently $S^* \neq \emptyset$.)

Proof. Find g and α from the separation theorem 2.2.3 with $K = S$. Since $0 \in S$, $\alpha \leqslant 0$. If there is $s_0 \in S$ with $g(s_0) < 0$, then since $\lambda s_0 \in S$ for each $\lambda \in \mathbb{R}_+$, there is λ such that $g(\lambda s_0) = \lambda g(s_0) < \alpha$. This contradiction shows that $g(S) \subset \mathbb{R}_+$, so $g \in S^*$; and $g(b) < \alpha \leqslant 0$.

Exercise. Let $S \subset X$ be a *closed* convex cone; use the separation theorem 2.2.3 to show that

$$x \in S \Leftrightarrow (\forall y \in S^*)\, y(x) \geqslant 0.$$

2.2.5 Cone inclusion theorem. Let $P \subset X$ and $Q \subset X$ be convex cones, with P closed. Then $P^* \subset Q^* \Rightarrow P \supset Q$.

Proof. If not, there is $q \in Q \backslash P$. Since P is a closed convex cone, the Cone Corollary 2.2.4 shows that there is $y \in P^*$ with $y(q) < 0$, so $y \in P^* \backslash Q^*$.

2.2.6 Farkas's theorem (for convex cones). Let X and Y be Banach spaces; $S \subset X$ a closed convex cone; $A \in L(X, Y)$; $C \in L(Y, X)$; $b \in Y$, $c \in Y'$. Then

(i) $$[A^T u \in S^* \Rightarrow u(b) \geqslant 0] \Leftrightarrow [b \in A(S)]$$

(assuming $A(S)$ is closed);

(ii) $$[Cv \in S \Rightarrow c(v) \geqslant 0] \Leftrightarrow [c \in C^T(S^*)]$$

(assuming $C^T(S^*)$ is weak $*$ closed).

Proof. Consider (i). Let $b = As$ for some $s \in S$; if $A^T u \in S^*$ then $0 \leqslant (A^T u)s = u(As) = u(b)$. Conversely, let $A^T u \in S^*$ imply $u(b) \geqslant 0$. Now

$$A^T u \in S^* \Leftrightarrow (\forall s \in S)(A^T u)s \geqslant 0$$
$$\Leftrightarrow (\forall s \in S)u(As) \geqslant 0 \Leftrightarrow u \in [A(S)]^*.$$

Similarly $u(b) \geqslant 0$ iff $u \in M^*$ where $M = \{\alpha b : \alpha \in \mathbb{R}_+\}$. Therefore $[A(S)]^* \subset M^*$. By the cone inclusion theorem 2.2.5, since $A(S)$ is closed, $A(S) \supset M$, so $b \in A(S)$.

Case (ii) is obtained by the replacement $X \mapsto X'$, $Y \mapsto Y'$ (for which the separation theorem 2.2.3 remains valid with a weak $*$ closed cone), and $S \mapsto S^*$, $S^* \mapsto S^\#$, where

$$S^\# = \{x \in X : (\forall s^* \in S^*)s^*(x) \geqslant 0\};$$

then $(S^*)^\# = S$.

2.2.7 Remarks. Farkas's original theorem was for finite systems of inequalities.

The hypothesis that the convex cone $A(S)$ (or $C^T(S^*)$) is closed is necessary; a three-dimensional example is given below, where Farkas's theorem is untrue without it. The cone is closed automatically if S (or S^*) is a *polyhedral cone* – the intersection of finitely many closed halfspaces whose boundaries contain zero – see the lemma 2.2.10. If $C(Y) = X$, then $C^T(S^*)$ can be proved closed; but it is then easier to prove Farkas's theorem directly. Let $Cy \in S$ imply $c(y) \geqslant 0$. Then

$$Cy = 0 \Rightarrow [Cy \in S \text{ and } C(-y) \in S]$$
$$\Rightarrow [c(y) \geqslant 0 \text{ and } c(-y) \geqslant 0] \Rightarrow c(y) = 0.$$

Define the linear map $h : X \to \mathbb{R}$ by $h(x) = c(y)$ whenever $y \in Y$ and $x = Cy$; h is uniquely defined since $C(Y) = X$ and $Cy = 0$ implies $c(y) = 0$. Then $c = hC$, and $h \in S^*$ since $x = Cy \in S$ implies $c(y) = h(Cy) \geqslant 0$; so $h \in S^*$, thus $c \in C^T(S^*)$. (See Appendix A.2 for proof of continuity of h.)

2.2.8 Example. Let S be the circular cone in $\mathbb{R}^3$, consisting of all vectors whose angle with the vector $(1 \quad 0 \quad 1)^T$ is $\leqslant \pi/4$. Then $(x \quad y \quad z) \in S$ iff

$$[1 \quad 0 \quad 1] \begin{bmatrix} x \\ y \\ z \end{bmatrix} \geqslant 2^{-1/2} \left\| \begin{bmatrix} 1 \\ 0 \\ 1 \end{bmatrix} \right\| \cdot \left\| \begin{bmatrix} x \\ y \\ z \end{bmatrix} \right\|; \text{ thus}$$

$$S = \left\{ \begin{bmatrix} x \\ y \\ z \end{bmatrix} \in \mathbb{R}^3 : x \geqslant 0, \ z \geqslant 0, \ 2xz \geqslant y^2 \right\}.$$

Now $P = \begin{bmatrix} 0 & 1 & 0 \\ 0 & 0 & 1 \end{bmatrix}$ is the orthogonal projection of $\mathbb{R}^3$ onto

the y, z plane. Although S is a closed convex cone, and P is a continuous linear map, the convex cone $P(S)$ is *not* closed. If $z > 0$, then for any y there is $x > 0$ such that $2zx \geqslant y^2$; but if $z = 0$, $y^2 = 0$, and if $z < 0$, there is no solution. So

$$P(S) = \left\{ \begin{bmatrix} y \\ z \end{bmatrix} \in \mathbb{R}^2 : 2xz \geqslant y^2, \ x \geqslant 0, \ z \geqslant 0 \right\}$$

$$= \left\{ \begin{bmatrix} y \\ z \end{bmatrix} \in \mathbb{R}^2 : z \geqslant 0 \right\} \cup \begin{bmatrix} 0 \\ 0 \end{bmatrix}.$$

In polar coordinates, setting $y = r \cos \theta$ and $z = r \sin \theta$, $P(S)$ is the sector $0 < \theta < \pi$.

2.2.9 Example. For this cone S, $S^* = S$. For $b^T = [1 \quad 0]$, $P^T v \in S^* \Rightarrow v_1 = 0 \Rightarrow b(v) = 0$, but $Pu = b$ has no solution $u \in S^3$.

2.2.10 Lemma. (on polyhedral cones). Let $B \in \mathbb{R}^{n \times k}$; then $S = B(\mathbb{R}_+^k)$ is a closed convex cone.

Proof. Clearly S is a convex cone. Let $\{y_p\}$ be a sequence in S, with $\{y_p\} \to y \in \mathbb{R}^n$. Then $y_p = Bx_p^+$ where $x_p^+ \in \mathbb{R}_+^k$. By reducing B to row echelon form, each solution x_p to $y_p = Bx_p$ has the form $x_p = Ay_p + z_p$, where A is a suitable matrix and $z_p \in B^{-1}(0)$. Let $z_p = n_p$ be the point of minimum norm in $B^{-1}(0)$ for which $\bar{x}_p \equiv Ay_p + n_p \in \mathbb{R}_+^k$; then n_p is a continuous function of y_p; since $\{y_p\} \to y$, the set $\{Ay_p : p = 1, 2, \ldots\}$ is compact, hence the sequence $\{n_p\}$ is bounded; hence $\{\bar{x}_p\}$ is bounded. Some subsequence of $\{\bar{x}_p\}$ converges, say to $\bar{x}$; then $\bar{x} \in \mathbb{R}_+^k$, and $B\bar{x} = \lim BAy_p = y$ since $Bn_p = 0$; thus $y \in S$. Hence S is closed.

* 2.2.11 Example. The separation theorem in $\mathbb{R}^n$ can be proved as follows. Let K be a closed convex subset of $\mathbb{R}^n$, and let $b \in \mathbb{R}^n \backslash K$. There is a sequence $\{q_r\} \subset K$ such that $\{\|b - q_r\|\} \to \delta \equiv d(b, K) > 0$. Some subsequence of $\{q_r\}$ converges, to $q \in K$; then $\|b - q\| = \delta$. Let $w \in K$; since K is convex, $w_\beta = q + \beta(w - q) \in K$ whenever $0 \leqslant \beta \leqslant 1$. Then $\delta^2 \leqslant \|b - w_\beta\|^2 = \|b - q\|^2 + \|q - w_\beta\|^2 + 2(w_\beta - q, q - b)$. Hence $\beta^2 \|q - b\|^2 + 2\beta(q - w, q - b) \geqslant 0$ for $0 \leqslant \beta \leqslant 1$. Hence $(q - w, q - b) \geqslant 0$ for each $w \in K$. The theorem follows, with $g = q - b$.

2.3 Critical points

Consider a function $f : X_0 \to \mathbb{R}$, and a constraint set $K \subset X$; assume that $K \subset X_0 \subset X$. The set $\mathbb{F} = \{f(x) : x \in K\}$ has, at the point $a \in K$,

a *global minimum* if $(\forall x \in K) f(x) \geqslant f(a)$;

a *local minimum* (or *minimum*) if, for some neighbourhood

$$N = \{x : \|x - a\| < \delta\}, (\forall x \in K \cap N) f(x) \geqslant f(a);$$

a *stationary point* if $|f(x) - f(a)|/\|x - a\| \to 0$ as $\|x - a\| \to 0$ with $x \in K$;

a global [local] *maximum* if $-\mathbb{F}$ has a global [local] minimum. Points a of these kinds are called *critical points* of $\mathbb{F}$, or of the constrained minimization problem

$$\text{Minimize } f(x) \text{ subject to } x \in K.$$

For a *constrained* minimization problem (thus $K \neq X$), critical points will often occur on the boundary ∂K (see 1.7). Note that $\mathbb{F}$ may have a finite infimum, without necessarily attaining a minimum.

2.4 Convex functions

A function $f : X \to \mathbb{R}$ is *convex* if, for all x, y and $0 < \lambda < 1$,

$$\lambda f(x) + (1 - \lambda) f(y) - f(\lambda x + (1 - \lambda) y) \in \mathbb{R}_+.$$

(Geometrically, each chord drawn on the graph of f lies *above* the graph.) A function f is *concave* if $-f$ is convex.

Exercise. Show that the real functions $f(x) = x^{2n}$ $(n = 1, 2, \ldots)$ are convex functions.

Exercise. Show that f is convex if its *epigraph* $\{(x, y) \in X \times \mathbb{R} : y \geqslant f(x)\}$ is a convex set.

A function $g : X \to Y$ is *S-convex*, where S is a convex cone in Y, if for all x, y and $0 < \lambda < 1$,

$$\lambda g(x) + (1 - \lambda)g(y) - g(\lambda x + (1 - \lambda)y) \in S.$$

Exercise. If $Y = \mathbb{R}^m$ and $S = \mathbb{R}^m_+$, show that g is S-convex iff each component of g is convex ($= \mathbb{R}_+$-convex).

Unless stated otherwise, the domain of each convex function is the whole space X.

2.4.1 Theorem. A local minimum of a convex function f on a convex set E is also a global minimum.

Proof. Let $f : E \to \mathbb{R}$ attain a local minimum at $p \in E$; assume that $f(x) < f(p)$ for some $x \in E$. Since E is convex, $\lambda x + (1 - \lambda)p \in E$ for each $\lambda \in (0, 1)$. Since f is convex,

$$f(p + \lambda(x - p)) - f(p) = f(\lambda x + (1 - \lambda)p) - f(p)$$
$$\leqslant \lambda f(x) + (1 - \lambda)f(p) - f(p) = \lambda[f(x) - f(p)] < 0$$

for arbitrarily small $\lambda > 0$, contradicting the local minimum.

Note that a *nonconvex* function may have several local minima, which are not global minima – e.g. $f(x) = x^2 + 4 \sin x$, $x \in \mathbb{R}$.

The following result characterizes convex functions in terms of derivatives; roughly, a function is convex if the chord lies above the tangent.

2.4.2 Theorem. Let $f : X \to Y$ be Fréchet differentiable; let $S \subset Y$ be a closed convex cone. Then f is S-convex iff, for all $x, z \in X$,

$$f(x) - f(z) - f'(z)(x - z) \in S. \qquad (\#)$$

* If $Y = \mathbb{R}$, and f is twice Fréchet differentiable, then f is convex iff $f''(z)$ is positive semidefinite, for each $z \in X$.

* **2.4.3 Remark.** If $Y = \mathbb{R}$, then f' maps X into $L(X, \mathbb{R}) = X'$, so $f''(x)$, the Fréchet derivative of f', is a linear map of X into X', or equivalently a map of $X \times X$ into $\mathbb{R}$ which is linear in each argument – write $f''(z)[u]^2$ if the arguments are the same. If $X = \mathbb{R}^n$ then $f''(z)[u]^2 = u^T M u$ where M is the $n \times n$ matrix representing $f''(z)$. Positive semidefinite means $f''(z)[u]^2 \geqslant 0$ for every u.

Proof. Let $x, z \in X$ and $0 < \lambda < 1$. If f is S-convex, then the convexity definition rearranges to

$$f(x) - f(z) - \lambda^{-1}[f(z + \lambda(x - z)) - f(z)] \in S.$$

Letting $\lambda \downarrow 0$ proves (#). Conversely, let (#) hold, let $\geqslant$ be the ordering induced by S, and denote $\xi = \lambda x + (1 - \lambda)z$, with $0 < \lambda < 1$; then

$$\lambda f(x) + (1 - \lambda)f(z) - f(\xi)$$

$$= \lambda[f(x) - f(\xi)] + (1 - \lambda)[f(z) - f(\xi)]$$

$$\geqslant \lambda f'(\xi)(1 - \lambda)(x - z) + (1 - \lambda)f'(\xi)\lambda(z - x)$$

$$= 0.$$

Let $Y = \mathbb{R}$, $S = \mathbb{R}_+$; let f be twice differentiable. If f is convex, then $f(x) - f(z) \geqslant f'(z)(x - z)$ and $f(z) - f(x) \geqslant f'(x)(x - z)$; hence $0 \leqslant [f'(x) - f'(z)](x - z) = f''(z)[x - z]^2 + o(\|x - z\|^2)$; hence $f''(z)$ is positive semi-definite. Conversely positive semidefiniteness gives, with $\xi = x + \lambda(z - x)$ and $\phi(\lambda) = f(\xi) - f'(z)(\xi - x)$, that, for some $\bar{\lambda} \in (0, 1)$ and $u \in [x, z]$,

$$f(x) - f(z) - f'(z)(x - z) = \phi(0) - \phi(1) = -\phi'(\bar{\lambda})(1 - 0)$$

$$= (1 - \bar{\lambda})f''(u)[z - x]^2 \geqslant 0.$$

2.4.4 Remark. Appendix A.7 extends the result (#) to $(\text{int } S)$-convex.

2.4.5 Theorem. If $E \subset \mathbb{R}^n$ is open, and $f : E \to \mathbb{R}$ is convex, then f is continuous on E.

* *Proof.* Let $e \in E$; then E contains an n-dimensional cube K with $e \in \text{int } K$; f is finite on each vertex of K; then convexity of f shows that f is bounded above on K, by the maximum of f on vertices of K. Let $e + h \in K$, and let $0 < \epsilon < 1$. Since f is convex,

$$(1 - \epsilon)f(e) + \epsilon f(e + h) \geqslant f(e + \epsilon h)$$

and
$$(1 + \epsilon)^{-1}[f(e + \epsilon h) + \epsilon f(e - h)] \geqslant f(e).$$

These rearrange to give

$$\epsilon[f(e) - f(e - h)] \leqslant f(e + \epsilon h) - f(e) \leqslant \epsilon[f(e + h) - f(e)]. \tag{*}$$

From the left inequality of (*), $-\frac{1}{2}f(e - h) \leqslant -\frac{3}{2}f(e) + f(e + \epsilon h)$; thus, since f is bounded above on K, it is also bounded below, say $|f(x)| \leqslant c$ for each $x \in K$. Then, from (*), $|f(e + \epsilon h) - f(e)| \leqslant 2\epsilon c$, which proves the continuity of f at e.

2.4.6 Remark. The discontinuous function f, given by $f(0) = f(1) = 1$, $f(x) = 0$ for $0 < x < 1$, is convex on the *closed* interval $[0, 1]$.

2.4.7 Lemma. If $g : X \to Y$ is S-convex, where $S \subset Y$ is a convex cone, then $K = \{x \in X : - g(x) \in S\}$ is a convex set.

Proof. Let $x, y \in K$, and $0 < \lambda < 1$. Then

$$-g(\lambda x + (1 - \lambda)y) = [-g(\lambda x + (1 - \lambda)y) + \lambda g(x)$$
$$+ (1 - \lambda)g(y)] + \lambda[-g(x)] + (1 - \lambda)[-g(y)]$$
$$\subset S + S + S \subset S.$$

2.4.8 Remark. A minimization problem is called *convex* if it has the form Minimize $f(x)$ subject to $x \in K$, where f is a convex function, and K is a convex set. If K is as in the lemma, then it is convex.

2.4.9 Remark. A theorem states that, if $E \subset X$ is a compact convex set, and f a continuous linear functional on E, then f attains the minimum (global, since E is convex) at a point of $\mathscr{C} \operatorname{extr} E$.

2.4.10 Remark. If $0 \neq p \in S^*$ and $s \in \operatorname{int} S$, then $s + N \subset S$ for some ball N, and $p(s) \geqslant 0$. If $p(s) = 0$ then $0 \leqslant p(s + N) = p(N)$; but $p(n) < 0$ for some $n \in N$. Hence $p(s) > 0$.

2.5 Alternative theorems

An *alternative theorem* states that two systems are so related that exactly one of the two has a solution. Much of the Lagrangean theory is built on such results. They all depend on the separation theorem for convex sets.

2.5.1 Basic alternative theorem. Let X and Y be real normed spaces; let $S \subset Y$ be a convex cone, with nonempty interior; let $\Gamma \subset X$ be convex; let $f : \Gamma \to Y$ be S-convex. Then exactly one of the two following systems has a solution:

(I): $\qquad\qquad -f(x) \in \operatorname{int} S, x \in \Gamma;$

(II): $\qquad\qquad (p \circ f)(\Gamma) \subset \mathbb{R}_+, 0 \neq p \in S^*.$

Proof. If both (I) and (II) have solutions, resp. x and p, then both $(p \circ f)(x) < 0$ and $(p \circ f)(x) \geqslant 0$, a contradiction. Suppose now that (I) has no solution.

The set $K = f(\Gamma) + \operatorname{int} S$ is open, since if $k = f(x) + s \in K$ ($x \in \Gamma, s \in \operatorname{int} S$), then $s + N \subset \operatorname{int} S$ for some ball N, so $k + N \subset K$. If (for $i = 1, 2$) $k_i = f(x_i) + s_i \in K$, and $0 < \tau < 1$, then

$$
\begin{aligned}
\xi \equiv (1 - \tau)k_1 + \tau k_2 &= (1 - \tau)(k_1 - f(x_1)) + \tau(k_2 - f(x_2)) \\
&+ [(1 - \tau)f(x_1) + \tau f(x_2) - f(\tau x_1 + (1 - \tau)x_2)] + f(x)
\end{aligned}
$$

$$
\text{where } x = \tau x_1 + (1 - \tau)x_2 \in \Gamma
$$

$$
\in \operatorname{int} S + \operatorname{int} S + S + f(x).
$$

So $\xi = s + f(x)$, where $s \in \operatorname{int} S$; so $\xi \in K$.

Thus K is an open convex set in Y, and $0 \notin K$ since (I) has no solution. By the separation theorem 2.2.3 there exists nonzero $p \in Y'$ such that $p(K) \subset \mathbb{R}_+$. Fix $x_0 \in \Gamma$. If $s \in \text{int} S$, then $s + N \subset \text{int} S$ for some ball N, and for λ large enough, $\lambda^{-1} f(x_0) \in N$, so $s - \lambda^{-1} f(x_0) \in \text{int} S$; since S is a cone, $\lambda s - f(x_0) \in \text{int} S$, so $\lambda s \in K$. Then $p(s) = \lambda^{-1} p(\lambda s) \geqslant 0$. Since p is continuous, it follows that $p(S) \subset \mathbb{R}_+$, thus $p \in S^*$. For any $x \in \Gamma$ and $e \in \text{int} S$, $k = f(x) + \epsilon e \in K$ for each $\epsilon > 0$, so that

$$(p \circ f)(x) = p(k) - \epsilon p(e) \geqslant 0 - \epsilon p(e) \to 0 \quad \text{as } \epsilon \downarrow 0.$$

2.5.2 Motzkin alternative theorem. Let X, Y, Z be normed spaces; let $S \subset Y$ be a convex cone, with $\text{int} S \neq \emptyset$; let $T \subset Z$ be a closed convex cone; let $A \in L(X, Z)$ and $B \in L(X, Y)$. If the convex cone $A^T(T^*)$ is weak $*$ closed, then exactly one of the two following systems has a solution:

(I'): $\qquad -Ax \in T, -Bx \in \text{int} S \ (x \in X)$;

(II'): $\qquad p \circ B + q \circ A = 0, q \in T^*, 0 \neq p \in S^*$.

Proof. With the substitutions $f = B$ and $\Gamma = -A^{-1}(T)$, (I') becomes $-f(x) \in \text{int} S, x \in \Gamma$. By theorem 2.5.1, this has no solution iff there is a solution $0 \neq p \in S^*$ to $(p \circ f)(\Gamma) \subset \mathbb{R}_+$, thus if $-Ax \in T$ implies $(p \circ B)(x) \in \mathbb{R}_+$. But this is equivalent, by Farkas's theorem since $A^T(T^*)$ is closed, to

$$(\exists p \in S^*, p \neq 0)(\exists q \in T^*) p \circ B = q \circ (-A),$$

which is (II').

2.5.3 Theorem. Let $\Gamma \subset X$ be convex; let $S \subset Y$ be a closed convex cone with $\text{int} S \neq \emptyset$; let $f : \Gamma \to \mathbb{R}$ and $g : \Gamma \to Y$ be resp. convex and S-convex functions. Then a necessary condition for the convex minimization problem:

(P_0): $\qquad$ Minimize $f(x)$ subject to $x \in \Gamma, -g(x) \in S$,

to attain a minimum at $x = a$, is that there exist $\tau \in \mathbb{R}_+$ and $v \in S^*$, not both zero, such that

$$(\forall x \in \Gamma)\ \tau(f(x) - f(a)) + v \circ g(x) \geqslant 0.$$

Proof. By the hypotheses, there is no solution to the system

$$-[f(x) - f(a)] \in \operatorname{int} \mathbb{R}_+, -g(x) \in S\ (x \in \Gamma).$$

Hence there is no solution $x \in \Gamma$ to the system

$$-([f(x) - f(a)], g(x)) \in \operatorname{int}(\mathbb{R}_+ \times S).$$

From the basic alternative theorem 2.5.1, there exists nonzero $p \in (\mathbb{R}_+ \times S)^*$, thus $p = (\tau, v) \in \mathbb{R}_+ \times S^*$, for which

$$p(f(x) - f(a), g(x)) = \tau(f(x) - f(a)) + v \circ g(x) \geqslant 0$$

for each $x \in \Gamma$.

Exercise. Deduce Farkas's theorem (2.2.6) from Motzkin's (2.5.2). (If $Cv \in S \Rightarrow c(v) \geqslant 0$, then there is no solution v to $Cv \in S$, $(-c)(v) \in \operatorname{int} \mathbb{R}_+$.)

Exercise. Show that either the linear system $Ax = c$ has a solution x, or the system $A^T v = 0$ has a solution v with $v(c) = 1$, but not both. (The first system is equivalent to $Ay - \tau c = 0$, $0 \neq \tau \in \mathbb{R}_+$. Then Motzkin's theorem 2.5.2 may be applied.)

Exercise. Motkin's theorem (2.5.2) still holds if the hypothesis that $A^T(T^*)$ is weak $*$ closed is replaced by $A(X) = Z$, or if T is a polyhedral cone.

2.6 Local solvability and linearization

In studying a nonlinear constrained minimization problem, it is often required to *linearize* the problem, that is, to approximate curved surfaces by tangent planes near a chosen point. Consider the problem:

(P): Minimize $f(x)$ subject to $-g(x) \in S$, $-h(x) \in T$.

Here X, Y, Z are Banach spaces; $X_0 \subset X$ is open; $f: X_0 \to \mathbb{R}$,

$g: X_0 \to Y$, $h: X_0 \to Z$ are Fréchet differentiable functions; $S \subset Y$ is a convex cone with int $S \neq \emptyset$; $T \subset Z$ is a closed convex cone. (The case $T = \{0\}$ is of particular interest.)

The procedure requires study of the solutions to $-h(x) \in T$ near a point $x = a$, assuming that $-h(a) \in T$. The system $-h(x) \in T$ will be called *locally solvable* at the point a if, for some $\delta > 0$, whenever the direction d satisfies $h(a) + h'(a)d \in -T$ and $\|d\| < \delta$, there exists a solution $x = a + \alpha d + \eta(\alpha)$ to $-h(x) \in T$, valid for sufficiently small $\alpha > 0$, where $\|\eta(\alpha)\|/\alpha \to 0$ as $\alpha \downarrow 0$. (For brevity, write $\eta(\alpha) = o(\alpha)$.) From the definition, a *linear* constraint $-h(x) \in T$ (one for which h is a linear function plus a constant) is automatically locally solvable. In appendix A.1, more general conditions are derived for local solvability, using the implicit function theorem. A *sufficient* condition is that h is continuously Fréchet differentiable, and that $h'(a)(X) = Z$.

Now associate to (P) the following linear system:

(PL): $-Aq \in T$, $-Bq \in \text{int}(\mathbb{R}_+ \times S)$ $(q \in X \times \mathbb{R})$,

in which $A = [h'(a)\ h(a)]$ and $B = \begin{bmatrix} f'(a) & 0 \\ g'(a) & g(a) \end{bmatrix}$ are continuous linear maps from $X \times \mathbb{R}$ to Z resp. Y.

2.6.1 Linearization theorem.
Let (P) and (PL) be as above; let the system $-h(x) \in T$ be locally solvable at the point a. If a is a local minimum for (P), then (PL) has no solution.

Proof. Suppose that (PL) has a solution $q = (d, \beta)$. For $\gamma > 0$, $h(a)\gamma + [h(a)\beta + h'(a)d] \in -T$. For sufficiently large γ, $d' = (\beta + \gamma)^{-1}d$ satisfies $\|d'\| < \delta$ and $h(a) + h'(a)d' \in -T$, so $-h(x) \in T$ has a solution $x = a + \alpha d + \eta(\alpha)$, where $\eta(\alpha)/\alpha \to 0$ as $\alpha \downarrow 0$. Then

$$f(a + \alpha d + \eta(\alpha)) - f(a) = \alpha f'(a)d + f'(a)\eta(\alpha) + \theta(\alpha),$$

where $\theta(\alpha)/\alpha \to 0$ as $\alpha \downarrow 0$. Since $f'(a)$ is continuous, and $-f'(a)d \in \text{int } \mathbb{R}_+$, $f(a + \alpha d + \eta(\alpha)) - f(a) < 0$ for sufficiently small α. However,

$$-g(a + \alpha d + \eta(\alpha)) = -g(a) - \alpha g'(a)d - g'(a)\eta(\alpha) - \rho(\alpha)$$

$$\text{where } \rho(\alpha)/\alpha \to 0 \text{ as } \alpha \downarrow 0$$

$$= (1 - \alpha\beta)[-g(a)] + \alpha[-g'(a)d - g(a)\beta] + \zeta(\alpha) \in S$$

as $\alpha \downarrow 0$, since $-g(a) \in S$, $-g'(a)d - g(a)\beta \in \text{int} S$, and $\zeta(\alpha)/\alpha \downarrow 0$ as $\alpha \downarrow 0$. So the minimum of (P) at a is contradicted.

This theorem is used in 4.4 to deduce Lagrangean necessary conditions for (P) to attain a minimum.

2.6.2 Example. Let $h : X \to \mathbb{R}^m$ be continuously differentiable, and let $-h(a) \in T = \mathbb{R}^m_+$. Then h partitions into vectors $h_{(1)}, h_{(2)}$ with $h_{(1)}(a) = 0$ and (each component of) $h_{(2)}(a) < 0$. Suppose that $h_{(1)}'(a)$ is an *onto* map. Let $h_{(1)}'(a)d = -s < 0$. Define $f(x, \lambda) = h_{(1)}(x) + \lambda s$, with real λ. Then $f(a, 0) = 0$, $f'(a, 0)$ maps *onto*, $[d \quad 1]^T \in (f'(a, 0))^{-1}(0)$, hence $f(x, \lambda) = 0$ has a solution $x = a + \alpha d + o(\alpha)$, $\lambda = 0 + \alpha 1 + o(\alpha)$; for sufficiently small $\alpha > 0$, $\lambda > 0$, so $h_{(1)}(x) = -\lambda s \leqslant 0$, and also $h_{(2)}(x) < 0$ since $h_{(2)}(a) < 0$. Hence $-h(x) \in \mathbb{R}^m_+$ is locally solvable at the point a.

References

Ben-Israel, A. (1969), Linear equations and inequalities in finite dimensional, real or complex, vector spaces: a unified theory, *J. Math. Anal. Appl.*, 27, 367–389. (For the counter example in 2.2.8)

Schaefer, H.H. (1966), *Topological Vector Spaces*, Macmillan, New York. (The separation theorem (2.2.3) is proved in Section II.9.)

Valentine, F.A. (1964, 1976), *Convex Sets*, McGraw-Hill; Krieger. (See Parts I and II for finite-dimensional convexity and separation theorems.)

CHAPTER THREE

Linear systems

3.1 Linear systems

Consider the *linear programming* problem

(LP):   Minimize $\{c^T v : Av \geqslant b, v \geqslant 0\}$.

where $v \in \mathbb{R}^n$, $c \in \mathbb{R}^n$, $b \in \mathbb{R}^m$, $A \in \mathbb{R}^{m \times n}$, and $\geqslant$ is taken componentwise. The constraint set Q is a closed polyhedron (the intersection of a finite number of closed halfspaces); since linear implies convex, any local minimum is a global minimum; if Q is nonempty and bounded, thus compact, then a minimum of (LP) is attained, at one of the finite number of extreme points of Q (see 2.1). Note that (LP) may be re-written with constraints $Cv \geqslant k$ $\left(\text{with } C = \begin{bmatrix} A \\ I \end{bmatrix}, k = \begin{bmatrix} b \\ 0 \end{bmatrix}\right)$.

Also, since $Av \geqslant b$ iff $Av - w = b$, where the *slack variable* $w \geqslant 0$, (LP) is equivalent to

(LP'):   Minimize $\{d^T x : Bx = b, x \geqslant 0\}$,

where $B = [A \mathbin{\vdots} -I]$ and $x^T = [v^T \mathbin{\vdots} w^T]$. If B has rank less than the number m of rows, then (LP') may be modified by adding suitable extra columns to B, to make the rank equal to m, and attaching to each additional component of x (called an *artificial variable*) a large positive component of d;

minimization will then eliminate all artificial variables, provided that the constraints $Bx = b$, $x \geqslant 0$ are consistent. Any $x \in Q$ is called a *feasible solution* of (LP').

Consider (LP') with rank of B equal to m, and $Q \neq \emptyset$ (i.e. consistent constraints). Suppose that Q is bounded. For each $x \in Q$, denote by B_x the matrix of those columns of B corresponding to positive components of x. Then, from the Lemma 2.1.2, $x \in \text{extr} \, Q$ iff B_x has linearly independent columns iff B has m linearly independent columns (called a *basis*), including those of B_x, iff $Bx = b$, $x \geqslant 0$ has a solution (called a *basic feasible solution*, b.f.s.) whose only positive components correspond to basis columns. Thus the bases correspond exactly to the extreme points of Q.

If Q is unbounded, then any $x \in Q$ has the form $x = u + \alpha h$ where $u \geqslant 0$, $\alpha \in \mathbb{R}$, $Bh = 0$. Since $x \geqslant 0$, either (i) $h \geqslant 0$, when ($\forall \alpha \geqslant 1$) $x = (\alpha - 1)h + (u + h) \geqslant 0$ so $x \in Q$, or (ii) $h \not\geqslant 0$, so $u + \alpha h \geqslant 0$ only for a finite interval of α. Then $d^T x = d^T u + \alpha d^T h$ is, for $x \in Q$, either *unbounded* below ((i) with $d^T h < 0$), or *bounded* below ((i) with $d^T h \geqslant 0$, or (ii)), when (LP') attains a minimum. Therefore *either $d^T Q$ is unbounded below, or* (LP') attains a minimum, say at $x = x_0$. Then this minimum is attained at an extreme point of Q, since an additional constraint $e^T x \leqslant \alpha$ could be added to bound the constraint set, without affecting the minimum. ($e^T = (1, 1, \ldots, 1)$.)

If $p \in \text{extr} \, Q$, but p is not a minimum for (LP'), then there is at least one edge of the polyhedron Q, joining p to another point of $\text{extr} \, Q$. The *simplex* algorithm (3.3) thus proceeds through a sequence of extreme points, until a minimum is found. Each extreme point is described by a basis; at each step (or *iteration*), one column of the basis is replaced by another column of B, so as to decrease $d^T x$. Unless a *degenerate* case occurs, where the decrease is zero, the process terminates, since the number of extreme points is finite. Degenerate cases where the calculation returns to some previous basis, and hence never terminates, exist in theory, but not in computational practice. Details of how to modify the simplex algorithm to avoid degeneracy are therefore omitted.

Exercise. If (LP)$'$ attains a minimum at $x = p$ and also at $x = q$, show that it also attains a minimum at any point on the line-segment $[p, q]$.

3.2 Lagrangean and duality theorems

For the problem

(LP$''$):   Minimize $\{c^T v : Cv \geqslant k\}$, $(C \in \mathbb{R}^{r \times n})$

define a Lagrangean function $L(x; m) = c^T x - m^T (Cx - k)$, where $m \in \mathbb{R}^r$.

3.2.1 Theorem. (LP$''$) attains a minimum at $x = a$ iff there exists $m \in \mathbb{R}^r_+$ such that

$$c^T - m^T C = 0 \text{ and } m^T (Ca - k) = 0. \qquad (*)$$

Remark. $c^T - m^T C$ is the derivative, at a, of L with respect to x.

Proof. Let (LP$''$) attain a minimum at $x = a$.

Let $u \in \mathbb{R}^n$ and $\beta \in \mathbb{R}$ satisfy $Cu + q\beta \in \mathbb{R}^r_+$, where $q = Ca - k$. Then either $u = 0$, when $c^T u \geqslant 0$ trivially, or $u \neq 0$, when for some $z \in \mathbb{R}^r_+$ and all sufficiently small $\alpha \in \mathbb{R}_+$,

$$C(a + \alpha u) - k = q + \alpha(z - \beta q) = (1 - \alpha\beta)q + \alpha z \in \mathbb{R}^r_+,$$

so $c^T(a + \alpha u) \geqslant c^T a$ since (LP$''$) has a minimum at a. Hence

$$[C \vdots q]\begin{bmatrix} u \\ \beta \end{bmatrix} \in \mathbb{R}^r_+ \Rightarrow [c^T \vdots 0]\begin{bmatrix} u \\ \beta \end{bmatrix} \geqslant 0.$$

By Farkas's theorem 2.2.6, noting that $[C \vdots q]^T (\mathbb{R}^r_+)$ is closed (by 2.2.10), there exist $m \in (\mathbb{R}^r_+)^* = \mathbb{R}^r_+$ such that

$$[c^T \vdots 0] = m^T [C \vdots q],$$

and this gives exactly $(*)$.

Conversely, if (∗) holds, and $Cx \geqslant k$, then

$$c^T x - c^T a = m^T(Cx - Ca) \geqslant m^T(k - Ca) = -m^T q = 0.$$

3.2.2 Duality. Given a problem (P), Minimize $\{f(v) : v \in Q\}$; a
problem (D), Maximize $\{\phi(z) : z \in N\}$, is called a *dual* of (P)
if there hold
(i) (*weak duality*) $f(v) \geqslant \phi(z)$ whenever $v \in Q$ and $z \in N$; and
(ii) if (P) attains a minimum at $v = a$,
then (D) attains a maximum at some point $z = z_0$, and
$f(a) = \phi(z_0)$; thus, the optima are equal.

Note that (D) is a dual of (P) if (i) holds, (P) attains a
minimum at $v = a$, and $\phi(z_1) = f(a)$ for *some* $z_1 \in N$; for then
z_1 is a maximum for (D). Note also that a dual of (D) may be
considered, by rewriting (D) as Minimize $\{-\phi(z) : z \in N\}$.

3.2.3 Theorem. (*Duality for linear programming*). The
problem

(LPD):   Maximize $\{k^T z : C^T z = c, z \geqslant 0\}$

is a dual of (LP″).

Proof. If $Cv \geqslant k$ and $C^T z = c, z \geqslant 0$, then $c^T v - k^T z = z^T(Cv - k) \geqslant 0$, so (i) holds. If (LP″) attains a minimum at
$v = a$, then, using the previous theorem, $c^T a - k^T m = m^T Ca - m^T k = 0$; also $m \geqslant 0$ and $C^T m = c$; so (ii) holds
(with $z_1 = m$).

3.2.4 Remark. If (LP″) does not attain a minimum, then $c^T v$
is unbounded below for (LP″); hence (i) implies that (LPD)
has no solution, i.e. the constraints for (LPD) are inconsistent.

Exercise. By expressing (LPD) in the equivalent form

$$\text{Minimize}\left\{ (-k)^T z : \begin{bmatrix} C^T \\ -C^T \\ I \end{bmatrix} z - \begin{bmatrix} c \\ -c \\ 0 \end{bmatrix} \geqslant 0 \right\},$$

obtain a dual of (LPD) as

$$
\text{Maximize} \left\{ [c^T \quad -c^T \quad 0] \begin{bmatrix} p \\ q \\ r \end{bmatrix} : [C \quad -C \quad I] \begin{bmatrix} p \\ q \\ r \end{bmatrix} = -k, \right.
$$

$$
\left. \begin{bmatrix} p \\ q \\ r \end{bmatrix} \geqslant 0 \right\};
$$

identify this with (LP″) by the substitution $v = q - p$, noting that $-Cv + r = -k$ and $r \geqslant 0 \Leftrightarrow Cv \geqslant k$. Thus, for *linear* programming, a dual of a dual gives the given (*primal*) problem. (This will *not* always be so for *nonlinear* problems.)

3.2.5 Remarks. Suppose that, in (LPD) and (LP″), k is *perturbed*, that is, changed by a small amount Δk. If Δk is small enough that the basis for (LPD) corresponding to the optimal solution is unchanged, then, by the Duality Theorem, the change in the optimal $c^T v$ equals the change in the optimal $k^T z$, namely $m^T \Delta k$.

If (LP″) represents a problem of economic optimization, then typically k is a vector of *requirements* which must be met by choice of v, and c is a vector of unit incremental costs. Then the vector m of *shadow costs* gives the change in *total* costs for small increments in the requirements. Whether or not the problem is one of economics, it is very important to analyse the sensitivity of a computed optimal solution to small change in the specification of the problem. (Neither data nor computation are completely precise.) Shadow costs are a (partial but useful) means for such a *sensitivity analysis*. (For a maximization problem, costs are replaced by prices; the interpretations are similar.)

3.3 The simplex method

An optimal solution of (LP′) can be computed by G.B. Dantzig's *simplex method* – named for an early application involving a simplex $\sum_{i=1}^{n} x_i \leqslant 1$, $(\forall i)x_i \geqslant 0$. Consider (LP′) in the form

$$\text{Minimize } \{px : x \geqslant 0, Bx = k\},$$

where $A \in \mathbb{R}^{m \times n}$, $x \in \mathbb{R}^n$, and p and k^T are row vectors. Assume that there is initially a b.f.s. (if necessary, add artificial variables so as to replace B by $[B \mid I]$, then the columns of I form a basis). To describe the algorithm, reorder the columns of B so that $x^T = [x^{(1)T} \mid x^{(2)T}]$, and correspondingly $B = [B^{(1)} \mid B^{(2)}]$, $p = [p^{(1)} \mid p^{(2)}]$. Then the initial feasible solution is $x^{(1)} = Qk$, $x^{(2)} = 0$, where $Q = B^{(1)-1}$. Denote the objective function by $f = px$.

Let x_j be an element of $x^{(2)}$, p_j the corresponding element of $p^{(2)}$, and b_j the corresponding column of $B^{(2)}$. If x_j is increased from 0, leaving all other elements of $x^{(2)}$ at 0, then $Bx = B^{(1)}x^{(1)} + B^{(2)}x^{(2)} = k$ and $f = p^{(1)}x^{(1)} + p^{(2)}x^{(2)}$ give

$$x^{(1)} = Qk - Qb_jx_j \text{ and } f = p^{(1)}Qk - (z_j - p_j)x_j,$$

where $z_j - p_j \equiv p^{(1)}Qb_j - p_j$. If $z_j - p_j > 0$, then f is decreased by increasing x_j; and x_j may increase from 0 up to the least positive value of $[Qk]_i/[Qb_j]_i$ $(i = 1, 2, \ldots, m)$, which occurs say at $i = s$. (A further increase of x_j would make $x_s < 0$, so the solution would no longer be feasible.) Giving x_j the value stated makes $x_s = 0$; so b_j must replace column s of $B^{(1)}$ in the basis, and p_j must replace element s of $p^{(1)}$. (It is usual, though not inevitable, to choose the largest positive z_j.)

Consider the matrices

$$D = \begin{bmatrix} -1 & \vdots & p^{(1)} \\ \cdots & \vdots & \cdots \\ 0 & \vdots & B^{(1)} \end{bmatrix} \text{ and } D^{-1} = H = \begin{bmatrix} -1 & \vdots & p^{(1)}Q \\ \cdots & \vdots & \cdots \\ 0 & \vdots & Q \end{bmatrix},$$

with rows and columns renumbered $0, 1, 2, \ldots, m$. If column s of D is replaced by $\begin{bmatrix} p_j \\ b_j \end{bmatrix}$, then H is premultiplied (see 1.8) by a matrix $I + N$, where N differs from the zero matrix only in column s; the element n_i in row $i \neq s$ is $-g_i/g_s$, and the element n_s in row s is $-1 + 1/g_s$, where

Table 3.1. Example of simplex calculation

Iteration	g	n	Matrix				Basis columns	$z_j - p_j$
1	3	−3/4	0	0	0	0	1, 5, 6 $j =$	j: 2 3 4
	−1	1/4	7	1	0	0		$z_j - p_j =$ −1 3 −2
	4	−5/4	12	0	1	0		Column 3 now replaces column 5 (see note below)
	3	−3/4	10	0	0	1		
2	1/2	−1/5	−9	0	−3/4	0	1, 3, 6	j: 2 4 5
	5/2	−3/5	10	1	1/4	0		1/2 −2 −3/4
	−1/2	1/5	3	0	1/4	0		Column 2 now replaces column 1
	−5/2	1	1	0	−3/4	1		
3	−11		−1/5	−4/5	0		2, 3, 6	j: 1 4 5
	4		3/5	1/10	0			−1/5 −12/5 −4/5
	5		1/5	3/10	0			All negative, so optimum has been reached.
	11		1	−1/2	1			

Table 3.1. Example of simplex calculation (continued)

Selection of column to leave the basis

Iteration	i	Values of $[Qk]_i / [Qb_j]_i$			s
1	3	7/(−1)	12/4	10/3	5
		(The least positive value, 12/4, corresponds to column 5)			
2	2	10/(5/2)	3/(−1/2)	1(−5/2)	1
		(The least positive value, 4, corresponds to column 1).			

$$g = H \begin{bmatrix} p_j \\ b_j \end{bmatrix} = \begin{bmatrix} z_j - p_j \\ Qb_j \end{bmatrix}.$$

Note that premultiplication of an m-rowed matrix M by $I + N$ changes M_{ij} to $M_{ij} + n_i M_{sj}$.

The data for (LP′) for a *data matrix* $\begin{bmatrix} 0 & \vdots & p \\ \hdashline k & \vdots & B \end{bmatrix}$. Premultiplication of this by H gives the *tableau*

$$\begin{bmatrix} p^{(1)}Qk & \vdots & 0 & \vdots & p^{(1)}QB^{(2)} - p^{(2)} \\ \hdashline Qk & \vdots & I & \vdots & QB^{(2)} \end{bmatrix}.$$

However, there is no need to apply the *iteration* (the replacement of one basis column by an appropriate new column) to the tableau; it suffices to iterate the matrix

$$W = \begin{bmatrix} p^{(1)}Qk & \vdots & p^{(1)} \\ \hdashline Qk & \vdots & Q \end{bmatrix},$$

which differs from H only in the initial column; and to compute the vector $p^{(1)}QB^{(2)} - p^{(2)}$ of $z_j - p_j$, for choosing the new basis column. The data matrix is generally *sparse* (contains a large proportion of zero elements), and is stored in any compact manner appropriate; the computer must store the matrix W (or equivalent information), and requires also the column g. (The tableau, although initially sparse, becomes less so after a number of iterations.) Note that, to maintain accuracy, the matrix W must be recalculated, by reinverting the relevant columns of the data matrix, after every so many (say m) iterations.

The minimum of f is obtained after a *finite* number of iterations, since the constraint set is a polyhedron with a finite number of vertices (see 2.1). This assumes that f decreases by a positive amount at each iteration. In a *degenerate* case, this decrease can be zero, and the algorithm may *cycle*, i.e. return to a basis previously considered, and so not terminate. This cycling can be avoided by slightly perturbing the problem, but this seems unnecessary in practice.

A simple example is as follows.

Minimize $x_2 - 3x_3 + 2x_4$ subject to $x_1, x_2, x_3, x_4 \geqslant 0$, and to

$$x_1 + 3x_2 - x_3 + 2x_4 = 7, \quad 2x_2 - 4x_3 \geqslant -12,$$
$$4x_2 - 3x_3 - 7x_4 \geqslant -10.$$

Introducing slack variables $x_5 \geqslant 0$ and $x_6 \geqslant 0$, and multiplying the last two constraints by -1, yields a data matrix

0	0	1	-3	2	0	0
7	1	3	-1	2	0	0
12	0	-2	4	0	1	0
10	0	-4	3	7	0	1
	↑				↑	↑

The columns marked ↑ form a unit submatrix, so no artificial variables are needed – a happy circumstance not found outside textbook examples. The iterations needed for optimizing this problem are given in Table 3.1 (Although serious use of the simplex method requires a computer, it is valuable to know what the computer is doing – especially since various algorithms for *nonlinear* problems use modifications of the simplex method.)

3.4 Some extensions of the simplex method

The simplex method has been extended to deal with *upper bounds* on variables, thus $0 \leqslant x_j \leqslant h_j$, without needing to include $x_j \leqslant h_j$ as an extra constraint.

The objective f may contain a parameter λ, say $p = u + \lambda v$, and the change of the solution is sought, as λ varies. (This is important for sensitivity analysis.) The simplex method is modified to carry two 'cost' rows, u and v; then, at a given iteration, $z_j - p_j$ has the form $\alpha_j + \lambda \beta_j$. Suppose that an optimal solution has been found, for given λ; then for each j not labelling a basis column, $\alpha_j + \lambda \beta_j \leqslant 0$. Hence the optimal basis just found remains an optimal basis for $\lambda_{min} \leqslant \lambda \leqslant \lambda_{max}$, where

$$\lambda_{min} = \max \{-\alpha_j/\beta_j : \beta_j < 0\}; \lambda_{max} = \min \{-\alpha_j/\beta_j : \beta_j > 0\}.$$

In this interval, the optimum value of f is a linear function of λ, namely $(u^{(1)} + \lambda v^{(1)})Qk$. When λ is increased to λ_{max}, the minimum defining λ_{max} is attained for some j, say j_0, and column j_0 now enters the basis, for $\lambda \geqslant \lambda_{max}$. A similar basis change occurs also when λ decreases below λ_{min}. (It is possible, in either case, for the linear program then to pass from bounded to unbounded objective.) A new interval of λ is now sought, in which the basis remains fixed; and so on. This method of *parametric programming* shows thus that the optimum value of the objective function is a *piecewise-linear* function of the parameter λ.

Procedures of similar kind can be used to find the dependence of the optimum on a parameter λ, occurring linearly in the requirement vector k; or when increase of λ not only increases p linearly, but introduces additional columns into B, as various values of λ are passed. The latter case arises in the *cutting stock* problem – that of finding the least wasteful way to cut up a roll of paper or a billet of steel into specified numbers of smaller pieces of specified size; here λ represents the size of the roll or billet supplied, and each column of B specifies a way of cutting it up; as λ increases, there are more such ways. The optimum value of the objective function is again a piecewise-linear function of λ.

A considerable class of nonlinear minimization problems can be approximated closely enough by problems similar to linear programming, to allow slight variants of the simplex method to solve them. A *separable programming* problem is a mathematical programming problem in which all nonlinear terms can be expressed as sums and differences of nonlinear functions of *single* variables; the nonlinearities are thus *separated*. There can be any number of linear terms as well; the method is most practicable when the nonlinear terms are few. Note that a product $x_1 x_2 = \frac{1}{4}[(x_1 + x_2)^2 - (x_1 - x_2)^2]$ can be expressed in *separated* form.

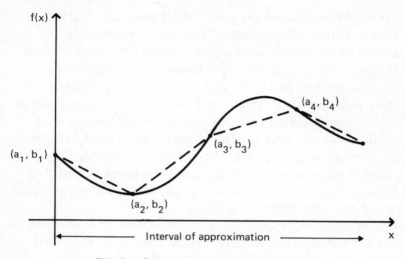

Fig. 3.1. Piecewise linear approximation

A continuous nonlinear function $f(x)$, where x is a single variable, can be approximated, over any finite interval, by a *piecewise-linear* function. Fig. 3.1 shows a simple example; in practice, more than five *breakpoints* (a_i, b_i) would be needed for adequate approximation. The approximating function is shown by a dashed line. Introduce non-negative variables λ_i, one for each breakpoint, satisfying the equations

$$\sum_{i=1}^{5} \lambda_i = 1; \quad \sum_{i=1}^{5} a_i\lambda_i = x; \quad \sum_{i=1}^{5} b_i\lambda_i = g(x);$$

where g is the approximating function to f, and $g(a_i) = b_i$ for each i. Thus if two *consecutive* λ_i are nonzero, and all the others zero, then the point represented lies in the segment of the graph indicated by the consecutive values of i. (If $\lambda_1 = \frac{1}{2} = \lambda_2$ and $\lambda_3 = 0 = \lambda_4$, the point is midway between (a_1, b_1) and (a_2, b_2).)

If the nonlinearities in the given problem are thus approximated by piecewise linear functions, then the approximated problem can be solved by a modified simplex method. For each *special ordered set* $\{\lambda_i\}$ as described above [see Beale (1970) for a more general account], at most two of the λ_i

can occur together in any basis, and if there are two, the indices i must be consecutive. This requirement limits the choice of the new column which can enter the basis, at each iteration. Each λ_i has an upper bound. The (modified) simplex method must converge, after a finite number of iterations, to a *local* (though not necessarily) *global* optimum of the approximated problem. It is necessary to start with a feasible solution near enough to the required local optimum (if there are more than one), to ensure convergence to it. The number of breakpoints along a nonlinear curve represents a balance between greater accuracy, and a larger linear program, which will take longer to solve.

A nonlinear objective function can always be exchanged for a linear objective function, at the cost of an additional nonlinear constraint. Thus it is equivalent to minimize $\{f(z) : -g(z) \in S\}$, or to minimize $\{t \in \mathbb{R} : t \geqslant f(z), -g(z) \in S\}$.

References

Beal, E.M.L. (1970), Advanced features for general mathematical programming, Chapter 4 of *Integer and Nonlinear Programming*, J. Abadie (Ed.), North-Holland.

Hadley, G. (1962), *Linear Programming*, Addison-Wesley, Reading. (For a general and detailed account of linear programming and the simplex algorithm.)

CHAPTER FOUR

Lagrangean theory

4.1 Lagrangean theory and duality

In this chapter, the Lagrangean and duality theorems for linear programming are extended to suitable nonlinear problems. Lagrangean *necessary* conditions for a constrained minimum are obtained, assuming either a convex problem (4.2), or differentiable functions (4.4); under some further restrictions, these Lagrangean conditions are also *sufficient* (4.5) to characterize a minimum. Two kinds of duality theory arise, one (4.2) depending only on convexity, the other (4.7) requiring also differentiable functions. The optimal dual variables can be interpreted as shadow costs, as in linear programming. Unlike linear programming, a dual of a nonlinear problem is not always unique; and the given (primal) problem is not always recovered as a dual of a dual (except under restrictions – see 4.8).

One example of a *convex* minimization problem is the *resource allocation problem* of 1.4, assuming that each f_i is a concave function, corresponding to a 'law of diminishing returns', that each X_i is a convex set, and that each $q_i : X_i \to \mathbb{R}^m$ is a convex function. The problem can then be written equivalently as

Minimize $f(x) = \sum_{i=1}^{k} [-f_i(x_i)]$ subject to $(\forall i)x_i \in X_i$, and

$$- \sum_{i=1}^{k} [q_i(x_i) - b/k] \in \mathbb{R}_+^m.$$

This problem has then the standard form

(P_0): Minimize $\{f(x) : x \in \Gamma, -g(x) \in S\}$,

where Γ is a convex subset of a normed vector space X, $f : \Gamma \to \mathbb{R}$ is a convex function, $g : \Gamma \to Y$ is an S-convex function, where S is a closed convex cone in the normed vector space Y. (In the example, $Y = \mathbb{R}^m$ and $S = \mathbb{R}_+^m$, so that each component of g is a convex function; and $\Gamma = X_1 \times X_2 \times \ldots \times X_k$.) To (P_0) corresponds a *Lagrangean* $f(x) + vg(x)$ $(v \in Y')$.

Exercise. If $X = \mathbb{R}^n$, $\Gamma = \{x \in \mathbb{R}^n : x^T A x \leqslant 1\}$, $f(x) = c^T x + (x^T B x)^{1/2}$, where A and B are positive semi-definite $n \times n$ matrices, and g is linear, show that (P_0) is a *convex* problem, i.e. both the constraint set and the objective function are convex.

The standard form for a *differentiable* minimization problem will be

(P_1):   Minimize$_{x \in X_0}$ $\{f(x) : -g(x) \in S, -h(x) \in T\}$.

Here X, Y, Z are Banach spaces; X_0 is an open subset of X; $S \subset Y$ is a convex cone, with $\mathrm{int}\, S \neq \emptyset$; $T \subset Z$ is a closed convex cone; and the functions $f : X_0 \to \mathbb{R}$, $g : X_0 \to Y$, and $h : X_0 \to Z$ are Fréchet differentiable. Usually T will be $\{0\}$; so that $h(x) = 0$, an equality constraint.

One example is the *quadratic programming* problem

(QP):   Minimize$_{x \in \mathbb{R}^n}$ $\{-c^T x + \frac{1}{2} x^T P x : Ax \leqslant b, \; x \geqslant 0\}$

where $c \in \mathbb{R}^n$, $P \in \mathbb{R}^{n \times n}$, $A \in \mathbb{R}^{m \times n}$, $b \in \mathbb{R}^m$. Note that (QP) is *convex* iff P is positive semidefinite. A more general quadratic programming problem (QP2) is obtained by adjoining quadratic constraints

$$m_i^T x + \tfrac{1}{2} x^T K_i x \leqslant d_i \ (i = 1, 2, \ldots, r) \quad \text{to} \quad (QP).$$

Another example of a differentiable problem is the optimal control problem of 1.6 and 1.10.

The problem (P_0) or (P_1) may be 'perturbed' by altering the constraint $-g(x) \in S$ to $-g(x) - z \in S$. If $-g(x) \in S$ takes the form $q(x) \geqslant b$, then the perturbation corresponds to adding an increment z to the 'requirements' vector b. The sensitivity of the problem to such a perturbation is closely connected with duality theory; *duality* here has the same definition as in 3.2.

4.2 Convex nondifferentiable problems

For the standard problem (P_0), define the *perturbation function*

$$F(z) = \inf \{f(x) : x \in \Gamma, -g(x) - z \in S\} \quad (z \in Y).$$

Denote by $(P_0(z))$ the perturbed problem occurring here. The problem

(D_0):   Maximize $\{\phi(v) : v \in S^*\}$,

where $\phi(v) = \inf \{f(x) + vg(x) : x \in \Gamma\} \ (v \in Y')$,

will turn out, under suitable hypotheses (see Theorem 4.2.13), to be a dual to (P_0).

4.2.1 Remark. Some *constraint qualification* (CQ) is needed, to ensure that the constraint set, near the minimum point, is not too complicated. *Karlin*'s CQ assumes that there is no nonzero $v \in S^*$ for which $vg(\Gamma) \subset \mathbb{R}_+$; *Slater*'s CQ assumes that there exists $x_0 \in \Gamma$ such that $-g(x_0) \in \text{int} \, S$; if $\text{int} \, S \neq \emptyset$, these two constraint qualifications are equivalent, by the basic alternative theorem (2.5.1).

4.2.2 Lemma. For the convex problem (P_0), $-\phi$ is convex on S^*, and F is convex.

Proof. Let $v, v' \in S^*$, and $0 < \lambda < 1$; then

$$\lambda\phi(v) + (1 - \lambda)\phi(v')$$

$$= \inf_{x \in \Gamma} \lambda(f(x) + vg(x)) + \inf_{x \in \Gamma} (1 - \lambda)(f(x) + v'g(x))$$

$$\leqslant \inf_{x \in \Gamma} \{\lambda(f(x) + vg(x)) + (1 - \lambda)(f(x) + v'g(x))\}$$

$$= \phi(\lambda v + (1 - \lambda)v').$$

Let $z_1, z_2 \in Y$, and let $\epsilon > 0$; let $z = \lambda z_1 + (1 - \lambda)z_2$, where $0 < \lambda < 1$. Then there are $x_i \in \Gamma$ ($i = 1, 2$) with $-g(x_i) - z_i \in S$ and $f(x_i) < F(z_i) + \epsilon$. Let $x = \lambda x_1 + (1 - \lambda)x_2$; since Γ and S are convex and g is S-convex, $x \in \Gamma$ and $-g(x) - z \in S$. Therefore

$$F(z) \leqslant f(x) \leqslant \lambda f(x_1) + (1 - \lambda)f(x_2)$$

$$< \lambda[F(z_1) + \epsilon] + (1 - \lambda)[F(z_2) + \epsilon].$$

Hence, since ϵ is arbitrary, $F(z) \leqslant \lambda F(z_1) + (1 - \lambda)F(z_2)$.

4.2.3 Remark. (D_0) is then equivalent to a *convex* minimization problem:

$$\text{Minimize } \{-\phi(v) : v \in S^*\}.$$

4.2.4 Saddlepoint theorem. (due to Kuhn and Tucker, Karlin, Uzawa). For the convex problem (P_0) to attain a minimum at $x = x_0 \in \Gamma$, it is *sufficient* that, for some $v_0 \in S^*$, the Lagrangean $\psi(x, v) = f(x) + vg(x)$ satisfies the saddlepoint condition

$$(\forall x \in \Gamma, \ \forall v \in S^*) \ \psi(x_0, v) \leqslant \psi(x_0, v_0) \leqslant \psi(x, v_0) \ (+)$$

This condition is also *necessary* if $\text{int } S \neq \emptyset$ and Karlin's CQ holds.

4.2.5 Remarks. Corresponding to the constraint inequality $-g(x) \in S$, the *Lagrange multiplier v* satisfies an inequality $v \in S^*$; if $S = \mathbb{R}^m_+$, then each component of v must be nonnegative; v is unrestricted only for an equality constraint, where $S = \{0\}$. The saddlepoint condition is sufficient, without convexity.

Proof. Assume (+). From the left inequality, ($\forall v \in S^*$) $(v - v_0)g(x_0) \leqslant 0$, so (if $v = 0$) $v_0 g(x_0) \geqslant 0$ and (if $v = v_0 + w$, for any $w \in S^*$) $wg(x_0) \leqslant 0$, hence $-g(x_0) \in S$. So $v_0 g(x_0) \leqslant 0$; but also $v_0 g(x_0) \geqslant 0$, so $v_0 g(x_0) = 0$. Hence, from the right inequality of (+), if $x \in \Gamma$, $-g(x) \in S$, then $f(x_0) + 0 \leqslant f(x) + v_0 g(x) \leqslant f(x)$, since $v \in S^*$. From this and $-g(x_0) \in S$, $x = x_0$ minimizes (P_0).

Assume that (P_0) is convex, int $S \neq \emptyset$, x_0 minimizes (P_0), and the Karlin CQ. Since (int $\mathbb{R}_+$) $\times S \supset \text{int}(\mathbb{R}_+ \times S)$, there is no solution $x \in \Gamma$ to

$$- [f(x) - f(x_0), g(x)] \in \text{int}[\mathbb{R}_+ \times S].$$

From the basic alternative theorem (2.5.1) there exists non-zero $(\tau_0, v_0) \in (\mathbb{R}_+ \times S)^*$, thus $\tau_0 \in \mathbb{R}_+$, $v_0 \in S^*$, not both zero, such that

$$(\forall x \in \Gamma) \quad \tau_0[f(x) - f(x_0)] + v_0 g(x) \geqslant 0.$$

Setting $x = x_0$, $v_0 g(x_0) \geqslant 0$; but $v_0 \in S^*$ and $-g(x) \in S$ give $v_0 g(x_0) \leqslant 0$, hence $v_0 g(x_0) = 0$. Also $v \in S^*$ and $-g(x_0) \in S$ imply $vg(x_0) \leqslant 0$. Hence (+) follows if $\tau_0 \neq 0$, for then $\tau_0 \neq 0$, for then $\tau_0 = 1$ can be assumed. But if $\tau_0 = 0$, then $0 \neq v_0 \in S^*$ and ($\forall x \in \Gamma$) $v_0 g(x) \geqslant 0$, contradicting Karlin's CQ.

4.2.6 Remark. The hypotheses of this theorem imply that $v_0 g(x_0) = 0$.

4.2.7 Definition. The problem $(P_0(z))$ is *stable* at $z = 0$ if, for some constant $k \in \mathbb{R}$, $F(z) - F(0) \geqslant k \|z\|$ for all sufficiently small $\|z\|$. The function F has a *subgradient* $w \in Y'$ at 0 if

$$F(z) \geqslant F(0) + wz \quad \text{for all sufficiently small } \|z\|.$$

Remarks. Stable means that the graph of F may not enter a cone, with vertex shifted to $(0, F(0))$ (except at the vertex). (Fig. 4.1 shows this, for $k < 0$ and $z \in \mathbb{R}$. If $k \geqslant 0$ then k may be replaced by any $k < 0$.) The *subgradient* is illustrated by the dashed line. (Note that F can change slope at 0 as shown.)

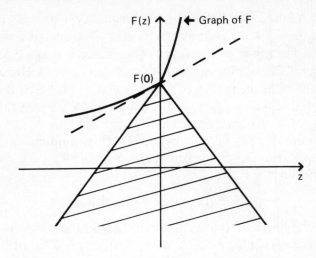

Fig. 4.1. Stability and subgradient

4.2.8 Theorem. The convex problem $(P_0(z))$ is stable at $z = 0$ iff F has a subgradient at 0.

Proof. If F has a subgradient, then $F(z) - F(0) \geqslant wz \geqslant -\|w\| \cdot \|z\|$, so $(P_0(z))$ is stable. Assume that $(P_0(z))$ is stable at $z = 0$; then

$$F(z) - F(0) \geqslant k\|z\| \qquad (*)$$

for sufficiently small $\|z\|$; it may be assumed that $k < 0$. But F is convex, therefore $F(\lambda z) - F(0) \leqslant \lambda[F(z) - F(0)]$ whenever $0 < \lambda < 1$. If, for some z_1, $F(z_1) - F(0) < k\|z_1\|$, then

$$F(\lambda z_1) - F(0) < \lambda k\|z_1\| = k\|\lambda z_1\|,$$

which contradicts $(*)$ for sufficiently small λ. Hence $(*)$ holds for all z. Therefore $A = \{(\alpha, z) \in \mathbb{R} \times Y : \alpha < k\|z\|\}$ is an open convex set, for $k < 0$; $B = \{(\alpha, z) \in \mathbb{R} \times Y : \alpha \geqslant F(z) - F(0)\}$ is convex, since F is convex; and $A \cap B = \emptyset$, by $(*)$. The separation theorem (2.2.3) then shows that there exist $\tau \in \mathbb{R}$, $w \in Y'$, not both zero, and $\beta \in \mathbb{R}$, with

$$\tau\alpha + wz \geqslant \beta \text{ on } B, \text{ and } \tau\alpha + wz \leqslant \beta \text{ on } A.$$

If $\tau < 0$ then $wz \leqslant \beta + (-\tau)\alpha$ on A, which is contradicted as $\alpha \downarrow -\infty$. If $\tau = 0$ then $wz \leqslant \beta$ for all $(\alpha, z) \in A$, and so for all $z \in Y$, contradicting $(\tau, w) \neq (0, 0)$. Hence $\tau > 0$. Hence, setting $\alpha = 0, z = 0$ in B, $\beta \leqslant 0$; letting $\alpha \uparrow 0$ and $\|z\| = 0$ in A, $\beta \geqslant 0$; so $\beta = 0$. Hence, choosing $\alpha = F(z) - F(0)$, $(\alpha, z) \in B$, hence, for each z,

$$F(z) - F(0) = \alpha \geqslant \tau^{-1}(\beta - wz) = (-w/\tau)z.$$

4.2.9 Weak duality theorem. If x satisfies the constraints of (P_0), and v satisfies the constraints of (D_0), then $f(x) \geqslant \phi(v)$.

Proof. Since $x \in \Gamma$, $-g(x) \in S$, and $v \in S^*$, $vg(x) \leqslant 0$, so

$$f(x) \geqslant f(x) + vg(x) \geqslant \phi(v).$$

4.2.10 Remarks. Hence $F(0) = \inf\{f(x) : x \in \Gamma, -g(x) \in S\} \geqslant \sup\{\phi(v) : v \in S^*\}$. Therefore (D_0) is a *dual* of (P_0) (see 3.2) under any hypothesis which ensures that the left and right sides of this inequality are equal, and that (D_0) attains its maximum.

4.2.11 Theorem. Let the convex problem (P_0) attain a finite minimum $F(0)$. Then (a) w is a subgradient of F at 0 iff (b) (D_0) attains its maximum at $v = w$, and $\text{Max}(D_0) = \text{Min}(P_0)$.

4.2.12 Corollary. $(P_0(z))$ is stable at $z = 0$ iff (D_0) is a *dual* of (P_0).

Proof. Assume (a); if $x \in \Gamma$, choose $z = -g(x)$, so that $-z - g(x) \in S$; then $F(0) \leqslant F(z) - wz \leqslant f(x) + wg(x)$. Hence

$$F(0) \leqslant \inf\{f(x) + wg(x) : x \in \Gamma\} = \phi(w).$$

If $w \notin S^*$, the separation theorem (2.2.3) shows that there is $b \in S$ with $wb = -1$. Suppose that the minimum $F(0)$ is attained at $x = a \in \Gamma$, $-g(a) \in S$. Then $(\forall \beta \geqslant 0)$ $F(-\beta b) - F(0) \geqslant w(-\beta b) = \beta$; and $F(0) = f(a)$. Since $b \in S$ and $-g(a) \in S$, $-g(a) - (-\beta b) \in S$ whenever $\beta \geqslant 0$. So, by definition of $F(-\beta b)$, $f(a) \geqslant F(-\beta b)$, thus

$F(-\beta b) - F(0) \leqslant 0$. Thus $0 \geqslant \beta$ for $\beta > 0$. Hence $w \in S^*$.
Then, by weak duality (4.2.9), $F(0) \geqslant \sup\{\phi(v): v \in S^*\}$; so
$v = w$ is optimal for (D$_0$), and $F(0) = \text{Min}(P_0) = \text{Max}(D_0)$.

Assume (b); then whenever $x \in \Gamma$ and $-g(x) - z \in S$,
$F(0) \leqslant f(x) + wg(x)$ and $w(g(x) + z) \leqslant 0$, so that
$F(0) \leqslant f(x) + wg(x) - w[g(x) + z] \leqslant f(x) - wz$. Hence

$$F(z) = \inf\{f(x): x \in \Gamma, -g(x) - z \in S\} \geqslant \text{Min}\{F(0) + wz\}$$
$$= F(0) + wz,$$

so w is a subgradient of F at 0.

4.2.13 Theorem. If Slater's CQ holds, then (D$_0$) is a dual of
(P$_0$).

Proof. Let the convex problem (P$_0$) attain a minimum at
$x_0 \in \Gamma$. Assume Slater's CQ; then Karlin's CQ follows from it,
by 4.2.1. Then the saddlepoint condition (+) is necessary for
a minimum of (P$_0$) at x_0, by the saddlepoint theorem (4.2.4);
also (+) implies that $v_0 g(x_0) = 0$, by 4.2.6. Hence, for each
$x \in \Gamma$,

$$f(x_0) = \psi(x_0, v_0) \leqslant \psi(x, v_0) = f(x) + v_0 g(x).$$

Consequently,

$$f(x_0) \leqslant \inf_{x \in \Gamma} [f(x) + v_0 g(x)] = \phi(v_0).$$

This, with weak duality (4.2.9), shows that $f(x_0) = \phi(v_0)$,
and that (D$_0$) attains a maximum at v_0. So duality follows.

4.2.14 Remark. Slater's CQ may be interpreted geometrically
as follows. Let B denote the set of $z \in Y$ for which
$-g(x) - z \in S$ for some $x \in \Gamma$. Slater's CQ assumes
$-g(x) \in \text{int } S$ for some $x \in \Gamma$, and hence $-g(x) - N \subset S$ for
some open ball N with centre $0 \in Y$; thus $0 \in \text{int } B$. The
function F is convex, by 4.2.2; if $Y = \mathbb{R}^n$, then F is continuous on N, by 2.4.5. It can then be shown that F has a subgradient w at 0, and hence (D$_0$) is a dual of (P$_0$), by 4.2.11.
[The separation theorem (2.2.3) finds a vector in $(Y \times \mathbb{R})'$

which separates the open convex set $\{(z, t) \in N \times \mathbb{R} : t > F(z)\}$ from $(0, F(0))$, and the subgradient follows from this.]

Exercise. Show that a linear program is *stable*.

4.3 Some applications of convex duality theory

Exercise. If (P_0) is the linear program:

$$\text{Minimize } c^T x \text{ subject to } Ax \geqslant b, \ x \geqslant 0,$$

the dual (D_0) is the same as that given by the theory of 3.2, namely

$$\text{Maximize } b^T v \text{ subject to } A^T v \leqslant c, \ v \geqslant 0.$$

Note that $\inf_{x \geqslant 0} [(c - A^T v)^T x + v^T b]$ is finite only when v satisfies $c - A^T v \geqslant 0$; so the latter constraint is necessary for the inf to attain its maximum with respect to v.

4.3.1 Remark. If (P_0) attains its minimum at $x = x_0$, and if duality holds, then $\text{Min}(P_0) = \inf_{x \in \Gamma} f(x) + wg(x)$, where $w \in S^*$ is the optimal value of v in (D_0).

Exercise. If F is (Fréchet) differentiable at 0, and w is a subgradient of F, then $w = F'(0)$. So if F is differentiable, and duality holds, then $wz \leqslant F(z) - F(0) = wz + o(\|z\|)$. Thus $w = F'(0)$ plays the role of a shadow cost, for the convex problem (P_0).

4.3.2 Example. The decentralized resource allocation problem of 1.4 and 4.1 is a convex problem, assuming the convexity hypotheses listed in 4.1. Then

$$\phi(v) = \inf \left\{ -\sum_i \{f_i(x_i) - v[q_i(x_i) - b/k]\} : (\forall i) x_i \in X_i \right\}$$

$$= -\sum_{i=1}^k [m_i(v) + vb/k],$$

where

$$m_i(v) = \sup\{f_i(x_i) - vq_i(x_i) : x_i \in X_i\} \quad (i = 1, 2, \dots, k),$$

and $v \in (\mathbb{R}_+^m)^* = \mathbb{R}_+^m$. Thus if ($D_0$) is a dual to ($P_0$), the given problem may be solved by solving the dual, and the latter reduces to minimizing $-\phi(v)$ over $v \in \mathbb{R}_+^m$, where, for each assumed value of the Lagrange multiplier v, $\phi(v)$ is obtained by solving k separate subproblems, each of them involving just one of the x_i. The vector v, which occurs as a parameter in each subproblem, may be interpreted as a *price* charged to the subsystem for resources used. If the optimal v is known, the subsystems are then optimized independently, and the system is thus *decentralized*. (See Arrow and Hurwicz, 1960.)

Exercise. Formulate Slater's CQ for this problem.

Exercise. The problem (QP) (see 4.1) is convex if P is positive semidefinite. Formulate Slater's CQ for (QP), and show directly that it implies Karlin's CQ, and also that duality holds. (Take $\Gamma = \mathbb{R}_+^m$.)

4.4 Differentiable problems

For the problem (P_1) from 4.1, namely

(P_1): $\underset{x \in X_0}{\text{Minimize}} \{f(x) : -g(x) \in S, \ -h(x) \in T\},$

with differentiable functions and $\text{int}\, S \neq \emptyset$, necessary conditions for a minimum can be given in terms of a modified Lagrangean

$$\tau f(x) + vg(x) + wh(x),$$

where $\tau \in \mathbb{R}_+$, $v \in S^*$, $w \in T^*$ are not all zero. The first (Fritz–John) theorem (4.4.1) allows that τ may be zero; the second (Kuhn–Tucker) theorem (4.4.3), under additional assumptions, has $\tau > 0$, and then (dividing by τ), τ can be replaced by 1. A counter example (4.4.6) shows that τ can be zero.

4.4.1 Theorem. For (P_1), let the constraint $-h(x) \in T$ be locally solvable at $a \in X_0$; let $[h'(a) \quad h(a)]^T (T^*)$ be weak $*$ closed. Then a necessary condition for (P_1) to attain a local minimum at $x = a$ is

(FJ): $\tau f'(a) + vg'(a) + wh'(a) = 0$; $vg(a) = 0$; $wh(a) = 0$;

$$\tau \in \mathbb{R}_+, \ v \in S^*, \ w \in T^*;$$

where τ and v are not both zero.

Proof. Let (P_1) attain a minimum at a; by the Linearization Theorem (2.6.1), the system $-Aq \in T, -Bq \in \text{int}(\mathbb{R}_+ \times S)$ has no solution, where

$$A = [h'(a) \ \vdots \ h(a)], \ B = \left[\begin{array}{c:c} f'(a) & 0 \\ \hdashline g'(a) & g(a) \end{array} \right].$$

Since the cone $A^T(T^*)$ is closed, Motzkin's alternative theorem (2.5.2) shows that there are $w \in T^*$ and nonzero $[\tau \quad v] \in (\mathbb{R}_+ \times S)^*$, such that $wA + [\tau \quad v]B = 0$; and this is exactly (FJ).

4.4.2 Remark. If $T = \{0\}$ (so $h(x) = 0$), and if $h'(a)(X)$ is a closed subspace of Z, and h is continuously differentiable, then either (a) $h'(a)(X) = Z$, then $-h(x) = 0$ is locally solvable (see Appendix A.1), and also Motzkin's theorem (2.5.2) applies, so that (FJ) follows with τ, v not both zero, or (b) $h'(a)(X) \neq Z$, then $h'(a)^T w = 0$ for some nonzero $w \in Z' = \{0\}^*$, hence (FJ) holds with this w, and $\tau = 0, v = 0$. (These are the hypotheses usually assumed.) The requirement that $\text{int} S \neq \emptyset$ can be weakened (Appendix A.4). Any *linear* constraint $-h(x) = -(Mx - b) \in T$ (where $M \in L(X, Z)$) is automatically locally solvable. The conditions $vg(a) = 0$ and $wh(a) = 0$ of (FJ) are called *complementary slackness* (or *transversality*) conditions.

Now consider the problem

(P_2):   Minimize $\{f(x) : -h(x) \in T\}$,
                           $x \in X_0$

obtained from (P_1) by omitting the constraint $-g(x) \in S$.

4.4.3 Theorem. For (P_2), let the constraint $-h(x) \in T$ be locally solvable at $a \in X_0$; let $[h'(a) \mid h(a)]^T (T^*)$ be weak $*$ closed. Then a necessary condition for (P_2) to attain a local minimum at $x = a$ is

(KT): $f'(a) + wh'(a) = 0, \ wh(a) = 0, \ w \in T^*.$

Proof. From the previous theorem (4.4.1), with $-g(x) \in S$ omitted, $\tau f'(a) + \bar{w}h'(a) = 0, \ \bar{w}h(a) = 0, \ \bar{w} \in T^*$, and $\tau \neq 0$ since $(\tau, v) \neq (0, 0)$ and v is absent. Then (KT) follows with $w = \bar{w}/\tau \in T^*$.

4.4.4 Remark. (FJ) and (KT) are called resp. the Fritz–John and the Kuhn–Tucker conditions. Observe that (FJ) assumes a local solvability hypothesis for only part of the constraint system.

The problem (P_2) is *regular* at $a \in X_0$ if, for some $\delta > 0$,

(Reg): $h(a) + h'(a)d \in -T$ and $\|d\| < \delta \Rightarrow f'(a)d \geqslant 0.$

Exercise. Show that local solvability at a minimum implies (Reg). (Note that, for a minimum,

$$f(a) \leqslant f'(a + \alpha d + o(\alpha)) = f(a) + f'(a)\alpha d + o(\alpha),$$

so $f'(a)d \geqslant 0$.)

4.4.5 Theorem. (KT) holds with the hypothesis of local solvability replaced by that of *regularity* at a.

Proof. Let $\alpha \in \mathbb{R}$ and $\xi \in X$ satisfy $\alpha h(a) + h'(a)\xi \in -T$. For any $\beta > 0$, since also $h(a) \in -T$, $\beta h(a) + \alpha h(a) + h'(a)\xi \in -T - T \subset -T$. Set $d = (\alpha + \beta)^{-1}\xi$, with β chosen so that $\alpha + \beta > 0$ and $\|d\| < \delta$; then $h(a) + h'(a)d \in -T$. If (Reg) is assumed, then $f'(a)d \geqslant 0$. Hence $f'(a)\xi \geqslant 0$. Thus

$$[h(a) \mid h'(a)]\begin{bmatrix} \alpha \\ \xi \end{bmatrix} \in -T \Rightarrow [0 \mid f'(a)]\begin{bmatrix} \alpha \\ \xi \end{bmatrix} \geqslant 0.$$

If $[h'(a) \mid h(a)]^T (T^*)$ is weak $*$ closed (or if $h'(a)(X) = Z$)

then Farkas's theorem (2.2.6) shows that there is $w \in T^*$ for which

$$[0 \mid f'(a)] = -w[h(a) \mid h'(a)].$$

Exercise. Show that (KT) implies (Reg); and that (Reg) holds for the quadratic programming problem (QP) given in 4.1.

4.4.6 Example. The problem

$$\text{Minimize } \{b^T x : P^T x \in S\},$$

where P, S, b are as in 2.2.8, has a minimum at $x = 0$; however (KT) does not hold there (see 2.2.9). Hence the 'closed cone' hypothesis cannot be omitted from the Kuhn–Tucker theorem 4.4.3. This hypothesis is fulfilled automatically if the cone T is polyhedral (see 2.2.10).

4.4.7 Remark. To see what (Reg) means, consider the particular case of constraints $g_i(x) \leq 0$ $(i = 1, 2, \ldots, m)$, $h_j(x) = 0$ $(j = 1, 2, \ldots, r)$. Suppose that $g_i(a) = 0$ for all $i \in J \subset \{1, 2, \ldots, m\}$, and $g_i(a) < 0$ for all $i \notin J$. Consider the condition

(Z): $[(\forall i \in J)g_i'(a)\xi \leq 0$ and $(\forall j)h_j'(a)\xi = 0] \Rightarrow f'(a)\xi \geq 0.$

Linearity shows that (Z) is unchanged by adjoining $\|\xi\| < \delta$ to the left side. Since $(\forall i \notin J)g_i(a) < 0$, δ can be chosen so that

$(\forall i \notin J)\|\xi\| < \delta \Rightarrow g_i(a) + g_i'(a)\xi \leq 0$; and $(\forall i \in J)g_i(a) = 0.$

Hence (Z) is equivalent to

(Z_1): $\qquad [(\forall i)g_i(a) + g_i'(a)\xi \leq 0, \ (\forall j)h_j'(a)\xi = 0,$

$$\|\xi\| < \delta] \Rightarrow f'(a)\xi \geq 0,$$

and thus to (Reg). So, for the present case, (Reg) is equivalent to the condition (Z), which involves only the *binding* constraints – those for which equality holds at a.

The *Kuhn–Tucker constraint qualification* (KTCQ) is the assumption that, whenever ξ satisfies the left side of (Z),

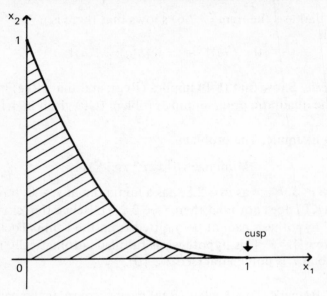

Fig. 4.2. Example where (KT) fails, after Kuhn and Tucker (1951)

there is a continuous arc $x = \omega(\alpha)$ $(\alpha \geqslant 0)$ satisfying the constraints, with $\omega(0) = a$ and initial slope $\omega'(0) = \xi$. The previous paragraph, and example 2.6.2, show that KTCQ is equivalent to local solvability of the constraint system at a. (Clearly KTCQ $\Rightarrow$ (Z).)

Exercise. Verify (Z) for linear constraints.

(Reg) requires that the boundary of the constraint set must be in some sense, smooth, near a. But a cusp is not always smooth enough, as the following example (from Kuhn and Tucker, 1951) shows.

4.4.8 Example. Consider the constraints:

$$-h(x) = \begin{bmatrix} (1 - x_1)^3 - x_2 \\ x_1 \\ x_2 \end{bmatrix} \geqslant \begin{bmatrix} 0 \\ 0 \\ 0 \end{bmatrix}$$

at the cusp $x_1 = 1$, $x_2 = 0$. At this point, $g_1(x) = g_3(x) = 0$,

$g_2(x) < 0$, and

$$g'(x) = \begin{bmatrix} 0 & 1 \\ -1 & 0 \\ 0 & -1 \end{bmatrix}.$$

So ξ satisfies the left side of (Z) iff

$$\begin{bmatrix} 0 & 1 \\ 0 & -1 \end{bmatrix} \begin{bmatrix} \xi_1 \\ \xi_2 \end{bmatrix} \geqslant \begin{bmatrix} 0 \\ 0 \end{bmatrix},$$

and thus ξ_1 is arbitrary, $\xi_2 = 0$. So the KTCQ is *not* fulfilled. If $f(x) = -x_1$, then $f(x)$ is minimized, over the constraint set, at the cusp. If (KT) holds, then $vg(a) = 0 \Rightarrow v_2 = 0$, contradicted by

$$f'(a) + vg'(a) = [-1 \ \vert \ 0] + [v_1 \ \vert \ v_2 \ \vert \ v_3] \begin{bmatrix} 0 & 1 \\ -1 & 0 \\ 0 & -1 \end{bmatrix}$$

$$= [0 \ \vert \ 0].$$

Note, however, that if $f(x) = -x_2$, then (KT) is satisfied with $v = [0 \ \vert \ 0 \ \vert \ 1]$; so KTCQ is *not* necessary for (KT).

This example is not a convex problem. However, Ben-Israel, Ben-Tal, and Zlobec (1976) have constructed a *convex* minimization problem where (KT) does not hold. A further example is as follows.

4.4.9 Example. Let K be a bounded closed convex set in $\mathbb{R}^n$; let $f(x) = cx$ $(c \neq 0)$ be a linear function which attains its minimum on K at a boundary point, a, of K; let $h(x)$ be the squared Euclidean distance from $x \in \mathbb{R}^n$ to K. Then $-h(x) \in \mathbb{R}_+$ iff $x \in K$. So $f(x)$ is minimized, subject to $-h(x) \in \mathbb{R}_+$, at $x = a$; but

$$f'(a) + vg'(a) = c + v(0) \neq 0$$

for any v; so (KT) does not hold.

Exercise. If the constraints are $-g(x) \in \mathbb{R}^m_+$, then (KT) follows if the rows of $g'(a)$ which correspond to binding constraints are linearly independent, and g is continuously differentiable. (See example 2.6.2.)

4.5 Sufficient Lagrangean conditions

4.5.1 Theorem. Let (FJ) hold with $\tau = 1$, f convex, g S-convex, and h affine (linear plus a constant). Then (P_1) is minimized at $x = a$.

Proof. $L(\cdot) = f(\cdot) + vg(\cdot) + wh(\cdot)$ is a convex function. If $-g(x) \in S$ and $-h(x) \in T$, then (FJ) gives $L'(a) = 0$ and $L(a) = f(a)$, so that

$$f(x) \geqslant L(x) \geqslant L(a) + L'(a)(x - a) = L(a) = f(a).$$

4.5.2 Theorem. Let (FJ) hold, with τ, v not both zero, f convex, g (int U)-convex, where $U \subset Y$ is a convex cone such that $S \subset U$ and $v \subset U^*$, and h T-convex (or affine). Then (P_1) is minimized at a.

4.5.3 Remarks. If int $S \neq \emptyset$, U may be taken as S. A function which is (int U)-convex is also called *strictly U-convex*; if $U = \mathbb{R}^m_+$, then each component of g satisfies a convexity inequality with *strict* inequality $>$ replacing $\geqslant$.

Proof. If $x = a$ is *not* a minimum for (P_1) then there is $x_0 \in X_0$ with $f(x_0) < f(a)$, $-g(x_0) \in S$, $-h(x_0) \in T$. Set $p = x_0 - a$. Since f is convex, $f(x_0) < f(a) \Rightarrow f'(a)p < 0$ by 2.4.2. Since g is (int U)-convex, 2.4.2 shows that

$$g(a) + g'(a)p$$

$$= -[g(x_0) - g(a) - g'(a)p] + g(x_0) \in - \text{int } U - U \subset - \text{int } U.$$

If $\tau > 0$, then $\tau f'(a)p < 0$; since $v \in U^*$ and $vg(a) = 0$ from (FJ), $vg'(a)p = v[g(a) + g'(a)p] \leqslant 0$, and is < 0 if $v \neq 0$; since $(\tau, v) \neq (0, 0)$, $[\tau f'(a) + vg'(a)]p < 0$. If h is T-convex (or affine), then similarly $wh'(a)p \leqslant 0$. Hence $[\tau f'(a) + vg'(a) + wh'(a)]p < 0$, contradicting (FJ).

*4.5.4 Remarks. This proof does not fully use the assumed convexity of f; only the weaker property, called *pseudo-convexity*, that

$$f(x) < f(a) \Rightarrow f'(a)(x - a) < 0,$$

is required. Also the hypotheses on g and h can be weakened. Let $b = g(a)$; let U_b denote the convex cone $\{\alpha(u - b) : \alpha \geqslant 0, u \in U\}$. Then $U \subset U_b$, and int $U_b = (\text{int } U)_b$. Now let g have the property

$$g(x) - g(a) \in - U_b \Rightarrow g'(a)(x - a) \in - \text{int } U_b. \qquad (*)$$

Assume also that h satisfies

$$h(x) - h(a) \in - T_{h(a)} \Rightarrow g'(a)(x - a) \in - \text{int } U_b.$$

(These properties are weaker consequences of the convexity hypotheses made in the above theorem.) If $x \neq a$ satisfies the constraints, then $-g(x) \in S$, hence $g(x) - g(a) \in -S_b \subset -U_b$, hence $g'(a)(x - a) \in - \text{int } U_b$. If $\tau = 0$, then $0 \neq v \in U^*$, hence $vg'(a)(x - a) < 0$; also (by a similar calculation for h) $wh'(a)(x - a) \leqslant 0$; and so $[vg'(a) + wh'(a)]v < 0$, contradicting (FJ). So the hypotheses imply that either there is only one point, a, satisfying the constraints (then a is trivially a minimum), or $\tau > 0$ (then (FJ) becomes (KT)).

Exercise. Complete the proof of a minimum, for the case $\tau > 0$.

4.6 Some applications of differentiable Lagrangean theory

The *quadratic programming* problem (QP) (see 4.1) has linear constraints, so a *necessary* condition for a minimum at $x = a$ is (KT): $-c^T + a^T P + \mu^T A - \lambda^T = 0$; $\mu^T (Aa - b) = 0$; $\lambda^T a = 0$; the Lagrange multipliers μ and λ satisfy $\mu \in \mathbb{R}_+^m$ and $\lambda \in \mathbb{R}_+^n$. From the constraints of (QP), $Aa - b \in - \mathbb{R}_+^m$ and $a \in \mathbb{R}_+^n$. If the problem (QP) is convex, thus if P is positive semidefinite, then these seven conditions are also *sufficient* for a minimum of (QP) at a.

Associate to (QP) the linear programming problem:

Minimize $e^T \xi + e^T y_-$ subject to

$$
\begin{bmatrix}
A & I & -I & 0 & 0 & 0 \\
P^T & 0 & 0 & A^T & -I & I
\end{bmatrix}
\begin{bmatrix}
x \\
y_+ \\
y_- \\
\mu \\
\lambda \\
\xi
\end{bmatrix}
=
\begin{bmatrix}
b \\
c
\end{bmatrix};
$$

$$x, y_+, y_-, \mu, \lambda, \xi \geqslant 0.$$

Here I is a unit matrix; e is a column of ones; ξ is an artificial variable; $b - Ax = y_+ - y_-$; and an initial basic feasible solution can be extracted from $y_+ - y_- = b$ and $-\lambda + \xi = c$, with other variables zero. The simplex method can then be applied, modified so that the columns of each basis correspond to at most one variable from each of the following pairs: $\{(y_+)_j, (y_-)_j\}, \{\lambda_j, x_j\}, \{\mu_j, (y_+)_j\}$. If P is positive definite, then this modified simplex method can be shown to converge to $\xi = 0, y_- = 0$; and the choice of bases makes $\mu^T y_+ = 0$ and $\lambda^T x = 0$. Thus the solution satisfies the seven conditions which characterize a minimum of (QP). Therefore (QP) is minimal at the value found for x.

This algorithm, called Wolfe's method (see Wolfe, 1959), can be extended to solve a quadratic program where P is positive semidefinite. However, the algorithm has the following disadvantages. If (as is typical) A is an $m \times n$ matrix with $n \gg m$, i.e. many more variables than constraints, then the simplex method for a L.P. problem with constraints need not consider matrices with more than $m + 1$ rows; but the matrix in Wolfe's method involves both A and A^T, so requires $m + n$ rows. A method more economical in matrix size would be preferable.

Consider the following *discrete-time optimal control problem*:

(DOC):   Minimize $J = \sum_{k=0}^{N} \psi(x_k, u_k, k)$

subject to the constraints

$$\Delta x_k \equiv x_{k+1} - x_k = \phi(x_k, u_k, k); -1 \leqslant u_k \leqslant 1$$
$$(k = 0, 1, 2, \ldots, N),$$

and to an *initial condition* $x_0 = c$, and to a *terminal constraint* $\zeta(x_{N+1}) = 0$. The given functions ψ, ϕ, ζ are assumed differentiable; c and N are constants; x_k and u_k may be real valued, or vector valued (then inequalities apply to each component). The *control function* $\{u_k\}_{k=0}^{N}$ must be chosen, in the region given by $-1 \leqslant u_k \leqslant 1$ $(k = 0, 1, \ldots, N)$ (the *control region*), and also the *trajectory* $\{x_k\}_{k=1}^{N}$, so as to minimize the *objective* J, while satisfying the constraints, including the initial condition and the terminal constraint.

Such a problem could arise in a context of inventory control (see 1.5), or as a discrete computational approximation to an optimal control problem in continuous time (1.6). A special case (DOC1) has $\phi(x_k, u_k, k) = a_k x_k + b_k u_k$, $\zeta(x) = x$, and

$$\psi(x_k, u_k, k) = \tfrac{1}{2} x_k^T A_k x_k + \tfrac{1}{2} u_k^T B_k u_k,$$

where x_k and u_k are column vectors (or, in particular, real variables), and a_k, b_k, A_k, B_k are constant matrices. Since (DOC1) is a quadratic programming problem, it could be solved, for example, by Wolfe's method. But the structure of (DOC) makes another method available. Note that the constraints of (DOC1) may be inconsistent – the point $x_{N+1} = 0$ is *not* necessarily *attainable* from the initial point $x_0 = c$. In what follows, the constraint set of (DOC) will be assumed consistent – so that there are *feasible* solutions; and it is required, among them, to find an *optimal* solution.

Assume, subject to verification, that a minimum is attained, and that a suitable constraint qualification is fulfilled there, so that (KT) holds. If $c \neq 0$, the initial condition $x_0 = c$ means that the space of $\{x_k\}$ in (DOC) is not a vector space. If (DOC) is reformulated in terms of $y_k = x_k - \epsilon_k$, with $\{\epsilon_k\}$ a fixed sequence with $\epsilon_0 = c$, then the theory leading to (KT) can be applied. The result is the same as that obtained, in the original variables x_k, in terms of the Lagrangean:

$$L = \sum_{k=0}^{N} \{\psi(x_k, u_k, k) + p_k(u_k - 1) - q_k(u_k + 1)$$
$$- \lambda_k[\Delta x_k - \phi(x_k, u_k, k)]\} + \sigma\zeta(x_{N+1}),$$

in which $p_k \geqslant 0$, $q_k \geqslant 0$, λ_k $(k = 0, 1, 2, \ldots, N)$ and σ are the Lagrange multipliers. (It is arbitrary whether the initial condition and the terminal constraint are included in the Lagrangean, or are excluded and treated separately as boundary conditions. To show what happens, the first has been excluded, and the second included.) Then (KT) gives:

(from $\partial L/\partial x_k$):

$$\psi_x(x_k, u_k, k) - \lambda_{k-1} + \lambda_k + \lambda_k \phi_x(x_k, u_k, k) = 0$$
$$(k = 1, \ldots, N),$$

(from $\partial L/\partial u_k$):

$$\psi_u(x_k, u_k, k) + \lambda_k \phi_u(x_k, u_k, k) + p_k - q_k = 0$$
$$(k = 0, \ldots, N),$$

(from $\partial L/\partial x_{N+1}$): $\sigma\zeta_x(x_{N+1}) - \lambda_N = 0$;

in which $\psi_x \equiv \partial\psi(x, u, k)/\partial x$, etc., and also the *transversality conditions* hold:

$$p_k(u_k - 1) = 0; \quad q_k(u_k + 1) = 0 \quad (k = 0, 1, \ldots, N).$$

The first equation of (KT) is a difference equation for $\Delta\lambda_{k-1}$. Setting $w_k = p_k - q_k$, the transversality conditions give

$$u_k \begin{cases} = +1 \\ = -1 \\ \in (-1, 1) \end{cases} \Rightarrow \begin{cases} q_k = 0 \\ p_k = 0 \\ p_k = q_k = 0 \end{cases} \Rightarrow w_k \begin{cases} \geqslant 0 \\ \leqslant 0 \\ = 0 \end{cases}.$$

In particular, for (DOC1), (KT) holds since the constraints are linear; and if A_k and B_k are positive semidefinite, then J is convex, so (KT) is also sufficient for a minimum. The first two equations of (KT) become $\Delta\lambda_{k-1} + x_k A_k + \lambda_k a_k = 0$

and $u_k B_k + \lambda_k b_k = -w_k$. The latter, with transversality, shows (if $B_k = 0$ and x_k, u_k are scalars) that $u_k = 1 \Rightarrow \lambda_k b_k \leqslant 0$, and $u_k = -1 \Rightarrow \lambda_k b_k \geqslant 0$. It will appear later that u_k need only take the boundary values ± 1; hence λ_k changes sign when u_k changes sign. The difference equations $\Delta x_k = a_k x_k + b_k u_k$ and $\Delta \lambda_{k-1} + x_k A_k + \lambda_k a_k = 0$ would then have to be solved, together with $\lambda_k b_k = -w_k$ and the initial and terminal conditions. This can be done step by step, starting with $x_0 = c$ and an assumed λ_0, which would later have to be adjusted to satisfy $x_{N+1} = 0$.

Exercise. Formulate (KT) for (QP) (in 4.1) with an additional quadratic constraint:

$$k^T x + \tfrac{1}{2} x^T K x \leqslant m.$$

Give a sufficient condition for (KT) to hold.
When is (FJ) sufficient for a minimum?

Exercise. Consider an inventory control model (compare 1.5):

Minimize $\sum_{t=1}^{n} [f(s_t) + b_t m_t]$ subject to $(\forall t) s_t \geqslant 0$, and

$\Delta s_t = A m_t - d_t$, $0 \leqslant m_t \leqslant k$ $(t = 0, 1, 2, \ldots, n-1)$.

Here the inventory vectors s_t form the 'trajectory'; the amounts m_t manufactured are the 'control', the demands d_t are given, A is a given matrix, f is convex, and b_t and k are constants. Formulate (KT), assuming that $s_0 = s_n = c$, a given vector. Note that this problem has the form (DOC), with the added constraints $s_t \geqslant 0$ (the system must not run out of stock); hence further terms are needed in the Lagrangean.

4.7 Duality for differentiable problems

Consider here the differentiable problem

(P_2): $\qquad$ Minimize $\{f(x) : -h(x) \in T\}$.
$\qquad\qquad\quad x \in X_0$

4.7.1 Duality theorem. In (P_2), let f be convex, let h be T-convex, let (P_2) attain a minimum at $a \in X_0$, and let (KT) hold at a. Then a dual to (P_2) is the problem

(D_2): $\underset{\xi \in X_0, w \in Z'}{\text{Maximize}} \{f(\xi) + wh(\xi) : w \in T^*, \; f'(\xi) + wh'(\xi) = 0\}.$

Exercise. If $-h(x) \in T$ has the form $-h_1(x) \in T_1$, $-h_2(x) = 0$, show that h_1 must be T_1-convex, and h_2 must be affine (i.e. $\{0\}$-convex).

Proof. Let $-h(x) \in T$ and let $w \in T^*$, $f'(\xi) + wh'(\xi) = 0$. Then

$f(x) - [f(\xi) + wh(\xi)] \geqslant f'(\xi)(x - \xi) - wh(\xi)$

$\qquad\qquad$ since f is convex (see 2.4.2)

$\qquad\qquad = -w[h(\xi) + h'(\xi)(x - \xi)]$

$\qquad\qquad$ substituting from the constraint of (D_2)

$\qquad\qquad \geqslant -wh(x)$ since wh is convex (by 2.4.2)

$\qquad\qquad \geqslant 0$ since $-h(x) \in T$ and $w \in T^*$.

This proves weak duality. Now, from (KT) for (P_2), there is $\bar{w} \in T^*$ with

$$f'(a) + \bar{w}h'(a) = 0 \text{ and } \bar{w}h(a) = 0;$$

so $(a, \bar{w})$ satisfies the constraints of (D_2), and

$$\text{Max}(D_2) \geqslant f(a) + \bar{w}h(a) = f(a) = \text{Min}(P_2).$$

This, with weak duality, shows that $(a, \bar{w})$ is optimal for (D_2).

Remark. This form of dual is called the *Wolfe dual*, after the originator.

Exercise. If the constraint of (P_2) is linear, thus $Ax - b \in K$, where K is a convex cone, then the Wolfe dual is

$\underset{\xi, w}{\text{Maximize}} \{f(\xi) - w(A\xi - b) : w \in K^*, \; f'(\xi) = wA\}.$

Exercise. If $f(\cdot)$ is linear, deduce the linear programming dual from this. (Only for a *linear* program can ξ be eliminated from the dual.)

Exercise. For (P_0), in 4.2, assume that the functions are differentiable as well as convex, and that Γ is an open set. Show then that the saddle point condition of 4.2.4 is equivalent to (KT). Discuss the relation between the Wolfe dual (4.7) and the convex dual (4.2) of this problem, assuming additionally that $f(x) + vg(x)$ attains its minimum over $x \in \Gamma$, for each $v \in S^*$.

Exercise. For the quadratic programming problem:

(QP2): $\underset{x \in \mathbb{R}^n}{\text{Minimize}} \{-c^T x + \tfrac{1}{2} x^T P x : Ax \leqslant b,\ x \geqslant 0,$

$$k^T x + \tfrac{1}{2} x^T K x \leqslant m\},$$

the Lagrangean is

$$L(x; \lambda, \mu, \rho) = -c^T x + \tfrac{1}{2} x^T P x + \lambda^T (Ax - b) - \mu^T x$$
$$+ \rho(k^T x + \tfrac{1}{2} x^T K x - m).$$

Then, if (QP2) is convex, its dual maximizes $L(\xi; \lambda, \mu, \rho)$ subject to $\partial L(\xi; \lambda, \mu, \rho)/\partial \xi = 0$ and $\lambda, \mu, \rho \geqslant 0$. Substituting for μ from the dual constraint gives an equivalent form of the dual objective as

$$\tilde{L} = -\tfrac{1}{2} x^T P x - \lambda^T b - \rho [\tfrac{1}{2} x^T K x + m].$$

(Extend this to several quadratic constraints.)

4.7.2 Example. Consider the problem:

$\underset{x_1, x_2, x_3 \in \mathbb{R}}{\text{Minimize}} \{9 - 8x_1 - 6x_2 - 4x_3 + 2x_1^2 + 2x_2^2 + x_3^2$

$+ 2x_1 x_2 + 2x_1 x_3 : x_1 \geqslant 0,\ x_2 \geqslant 0,\ x_3 \geqslant 0,$

$x_1 + x_2 + 2x_3 \leqslant 3\}.$

This problem is convex, since the matrix of quadratic coefficients:

$$\begin{bmatrix} 2 & 1 & 1 \\ 1 & 2 & 0 \\ 1 & 0 & 1 \end{bmatrix}$$

is positive definite. The optimal solution is 1/9, at (4/3, 7/9, 4/9). The dual constraints, obtained by differentiating the Lagrangean, reduce to:

$$-8 + 4x_1 + 2x_2 + 2x_3 + \lambda \geqslant 0, \quad -6 + 4x_2 + 2x_1 + \lambda \geqslant 0,$$
$$-4 + 2x_3 + 2x_1 + 2\lambda \geqslant 0, \quad \lambda \geqslant 0; \text{ and}$$
$$\tilde{L} = 9 - [2x_1{}^2 + 2x_2{}^2 + x_3{}^2 + 2x_1 x_2 + 2x_1 x_3] - 3\lambda.$$

(Here $x = (x_1, x_2, x_3)$ is written for the dual variable also.)

Table 4.1.

(x_1, x_2, x_3)	Objective of given problem	$\tilde{L}$	Dual constraint	Objective of dual
1, 1, 1/2	1/4	$7/4 - 3\lambda$	$\lambda \geqslant 1$	$-5/4$
3/2, 3/4, 3/8	9/64	$-9/64 - 3\lambda$	$\lambda \geqslant 1/8$	$-33/64$
1, 1, 1	Not feasible	$0 - 3\lambda$	$\lambda \geqslant 0$	0

Note that (x, λ) can be feasible for the dual, without x being feasible for the given program. Note also that a good approximation to the optimal objective does *not* imply a good approximation to the vector where the optimum occurs, since the gradient may be small near an optimum.

Beale (1967) gives the worked solution of this problem by *Beale*'s method (see also Section 7.6).

Since the dual is *linear* in the multiplier λ (this holds generally, not only for quadratic programs), the dual can be maximized with respect to λ by a linear program, if a value of x is assumed. By weak duality (valid even without verifying that duality holds), this maximum gives a *lower bound* to the optimal value of the objective in the given problem. Since an optimum of a nonlinear program can usually only be obtained by a (hopefully convergent) sequence of iterations, such a lower bound may be computationally useful. (This was proposed by Hanson (1961), in particular for a quadratic problem with quadratic constraints.) If duality holds, this bound can be close to the optimum.

For the above simple example, some instances are given in Table 4.1.

Exercise. The nonlinear constraint, in $\mathbb{R}^3$, given by

$$b_3 - (Ax)_3 \geqslant c\{\alpha[b_1 - (Ax)_1]^2 + \beta[b_2 - (Ax)_2]^2\}^{1/2},$$

where $c, \alpha, \beta \in \mathbb{R}_+ ; b \in \mathbb{R}^3 ; A \in \mathbb{R}^{3 \times r}$; is equivalent to the *linear* constraint $-(Ab - b) \in S$, where S is the elliptical cone

$$S = \{z \in \mathbb{R}^3 : z_3 \geqslant c[\alpha z_1{}^2 + \beta z_2{}^2]^{1/2}\}.$$

4.8 Converse duality

Under suitable conditions on the mathematical programming problem P, there exists another problem D which is a dual to P. Often there exists D which can be calculated from P by some systematic rule, which may be denoted $D = F(P)$ – e.g. 4.2 and 4.7. A *converse duality* theorem gives conditions under which F applies also to D, and $F(D) = P$. This is true for linear programs (2.2), and for the quadratic program (QP) in 4.1, assuming P is positive semidefinite. (Exercise. Prove this.) Aside from these cases, the known converse duality theorems require the nonlinear terms to be dominant; no such results seem to exist for slightly nonlinear problems.

For (P_2) (see 4.7), form the Lagrangean $L(x, w) = f(x) + wh(x)$ $(x \in X_0, w \in T^*)$. The Fréchet derivative H of $L'(x, w) \equiv \partial L(x, w)/\partial x$ with respect to (x, w) at $(a, \bar{w})$ is $[M \mid h'(a)^T]$, where $M = f''(a) + \bar{w}h''(a)$, assuming that these second Fréchet derivatives exist. As in 2.3, $f''(a)$ and M will be regarded as continuous linear maps from X into X'.

4.8.1 Converse duality theorem. In (P_2), let f be convex, and g T-convex; let f and g be twice continuously Fréchet differentiable; let M be a bijection of the Banach space X onto its dual space X'; let (D_2) attain a local maximum at $(\xi, w) = (a, \bar{w})$. Then (P_2) is a dual of (D_2).

Remark. The requirement that M is a bijection becomes, if $X = \mathbb{R}^n$, the requirement that the $n \times n$ matrix M is nonsingular.

Proof. Weak duality is proved as in 4.7.1. If the hypotheses imply that $-h(a) \in T$ and $\bar{w}h(a) = 0$, then $x = a$ satisfies the constraints of (P_2), and $\text{Min}(P_2) \leqslant f(a) = f(a) + \bar{w}h(a) = \text{Max}(D_2)$; this, with weak duality, shows that (P_2) attains its minimum at a, and hence (P_2) is a dual of (D_2).

Since $L'(a, \bar{w}) = f'(a) + \bar{w}h'(a) = 0$ from the dual constraint, L' is continuously Fréchet differentiable, and $M = \partial L'(a, \bar{w})/\partial x$ is bijective by hypothesis, the implicit function theorem (see Appendix A.1), shows that, whenever $w = \bar{w} + t$ and $\|t\|$ is sufficiently small, the equation $L'(\xi, w) = 0$ has a solution $\xi = \xi(t)$, for which $\xi(t) \to a$ as $t \to 0$. Choose t so that $\bar{w} + \alpha t \in T^*$ for $0 < \alpha \leqslant 1$; then $(\xi(\alpha t), \bar{w} + \alpha t)$ satisfies the constraints of (D_2). Since (D_2) is maximized at $(a, \bar{w})$,

$$0 \geqslant [f(a + \xi(t)) + (\bar{w} + t)h(a + \xi(t))] - [f(a) + \bar{w}h(a)]$$
$$\geqslant f'(a)\xi(t) + \bar{w}h'(a)\xi(t) + th(a + \xi(t))$$

$$\text{by the convexity (2.4.2)}$$

$$= th(a + \xi(t)) \qquad \text{by (FJ).}$$

Replacing t by αt, then letting $\alpha \downarrow 0$, shows that $th(a) \leqslant 0$.

Consequently $th(a) \leqslant 0$ whenever $t \in T^*$, hence $-h(a) \in T$. Also, setting $t = -\frac{1}{2}\bar{w}$, $\bar{w} + \alpha t \in T^*$ for $0 < \alpha \leqslant 1$, hence $(-\frac{1}{2}\bar{w})h(a) \leqslant 0$. But $\bar{w}h(a) \leqslant 0$ since $\bar{w} \in T^*$, and $-h(a) \in T$ has been proved. Hence $-h(a) \in T$ and $\bar{w}h(a) = 0$, as required.

4.8.2 Remarks. A constraint qualification has *not* been assumed here for (P_2); although the requirement that M is bijective may be regarded as a constraint qualification. Convexity is required to obtain weak duality. Although (P_2) is a convex problem, (D_2) is not (except in the linear programming case).

Exercise. Apply this theorem to quadratic programming with quadratic constraints.

Exercise. Apply the converse duality theorem to the optimal control problem of 1.10.

References

Arrow, J.J., and Hurwicz, L. (1960), Decentralization and computation in resource allocation, in *Essays in Economics and Econometrics*, University of North Carolina Press.

Beale, E.M.L. (1967), Numerical Methods, in J. Abadie (Ed.), *Nonlinear Programming*, North-Holland. (This reference contains the worked solution of the problem 4.7.2 by *Beale's method*.)

Ben-Israel, A., Ben-Tal, A., and Zlobec, S. (1976), Optimality conditions in convex programming, IX International Symposium on Mathematical Programming, Budapest.

Craven, B.D., and Mond, B. (1971), On converse duality in nonlinear programming, *Operations Research*, **19**, 1075–1078. (Some other references are listed in this paper.)

Craven, B.D., and Mond, B. (1973), Transposition theorems for cone-convex functions, *SIAM J. Appl. Math.*, **24**, 603–612. (For the Fritz–John theorem in Banach space, and an optimal control application.)

Craven, B.D. (1975), Converse duality in Banach space, *J. Optim. Theor. Appl.*, **17**, 229–238.

Craven, B.D., and Mond, B. (1978), Lagrangean conditions for quasi-differentiable optimization, *Proceedings of the IX International Symposium on Mathematical Programming* (Budapest, 1976), North-Holland, in the press. (This paper includes a converse duality theorem.)

Hanson, M.A. (1961), A duality theorem in non-linear programming with non-linear constraints, *Australian Journal of Statistics*, **3**, 64–72.

Kuhn, H.W., and Tucker, A.W. (1951), Nonlinear programming, in *Proceedings of the Second Berkeley Symposium on Mathematical Statistics and Probability*, J. Neyman (Ed.), University of California Press, Berkeley, 481–492.

Wolfe, P. (1959), The simplex method for quadratic programming, *Econometrica*, **27**, 382–398.

CHAPTER FIVE

Pontryagin theory

5.1 Introduction

For the discrete optimal control problem (DOC1) of 4.6, define the function

$$H(x, u, \lambda) = \sum_{k=0}^{N} [\psi(x_k, u_k, k) + \lambda_k \phi(x_k, u_k, k)]$$

$$= \sum_{k=0}^{N} [\tfrac{1}{2} x_k^T A_k x_k + \tfrac{1}{2} u_k^T B_k u_k + \lambda_k (a_k x_k + b_k u_k)]$$

$$\equiv \sum_k h_k(x_k, u_k, \lambda_k)$$

where x is the column vector of $x_0, x_1, \ldots, x_N$, and similarly u of $u_0, \ldots, u_N$ and λ of $\lambda_0, \ldots, \lambda_N$. Suppose that the minimum of (DOC1) is reached at $(x, u, \lambda) = (\xi, \eta, \tilde{\lambda})$. For the associated problem

$$\text{Minimize}_u \ H(\xi, u, \tilde{\lambda}) \ \text{subject to} \ -1 \leqslant u_k \leqslant 1$$
$$(k = 0, 1, \ldots, N),$$

the constraints are linear, hence a *necessary* condition for a minimum at $u = \eta$ is (KT), namely

$$\partial H(\xi, u, \tilde{\lambda}) / \partial u_k \, |_{u = \eta} = -p_k + q_k \quad (k = 0, 1, \ldots, N),$$

$$(\#)$$

where $p_k \geqslant 0$ and $q_k \geqslant 0$ satisfy the same transversality conditions as previously obtained in 4.6. But since

$$\partial H(\xi, u, \tilde{\lambda})/\partial u_k \big|_{u=\eta} = \psi_u(\xi_k, \eta_k, k) + \lambda_k \phi_u(\xi_k, \eta_k, k)$$

$$= \sum_{k=0}^{N} [\eta_k B_k + \lambda_k b_k],$$

(#) is exactly the second equation of (KT) for (DOC1). If H is a convex function of u, thus if B_k is positive semidefinite for each k, then (KT) for the associated problem is also *sufficient* (see 4.5) for a minimum.

Hence (KT) for (DOC1) holds, assuming B_k is positive semidefinite, if and only if both

(i) $$\Delta\lambda_{k-1} + x_k^T A_k + \lambda_k a_k = 0,$$

and (ii) the associated problem attains a minimum at $u = \eta$. But (KT) for (DOC1) is also *sufficient* for a minimum of (DOC1) if each A_k and B_k is positive semidefinite. Assuming this, (i) and (ii) are necessary and sufficient conditions for a minimum of (DOC1). Now (ii) holds iff, for each k, $h_k(\xi_k, u_k, \tilde{\lambda}_k)$ is minimized with respect to u_k in $-1 \leqslant u_k \leqslant 1$. This is an instance of *Pontryagin's Principle*. (Pontryagin[†] considered continuous-time problems, but the ideas apply also to discrete time; the result is usually called the Maximum Principle, corresponding to maximizing $-h_k$.)

For (DOC1), if each $B_k = 0$, then h_k is linear in u_k, and the constraints are linear, hence the optimum is reached when each u_k is either $+1$ or -1. Hence the optimal control vector $u = \{u_k\}_{k=0}^{N}$ assumes only values on the *boundary* of the control region $\{u : (\forall k) - 1 \leqslant u_k \leqslant 1\}$, and jumps between them; this behaviour is called *bang-bang control*.

To construct a more general theory, a preliminary result is needed. Let $H : U \to \mathbb{R}$ and $G : U \to Q$ be continuously Fréchet differentiable functions, let U and Q be Banach spaces, and let $K \subset Q$ be a convex cone. The problem

$$\text{Minimize}_{u \in U} H(u) \text{ subject to } G(u) \in K \qquad (\#)$$

[†] Three syllables: Pon-trya-gin (y is a consonant).

has the *quasimin* property (QM) at $u = \eta$ if $G(\eta) \in K$, and for some function $\theta(u) = o(\|u - \eta\|)$,

$$H(u) - H(\eta) \overset{\wedge}{+} \theta(u) \geqslant 0 \quad \text{for} \quad \|u - \eta\| \to 0, \quad G(u) \in K.$$

Since $\theta(u) = o(\|u - \eta\|)$ means that $\|\theta(u)\| < \epsilon\|u - \eta\|$ whenever $\|u - \eta\| < \delta(\epsilon)$, (QM) occurs iff

$$\lim \inf_{u \to \eta, G(u) \in K} [H(u) - H(\eta)]/\|u - \eta\| \geqslant 0.$$

Exercise. If $\{u_n\} \to \eta$, $(\forall n)G(u_n) \in K$, show that $H(u_n) - H(\eta) \geqslant \beta_n$, with $\{\beta_n\} \to 0$.

Exercise. Show that (QM) *for a convex problem* implies a minimum. (This is not true without convexity; consider $H(u) = -u^2, u \in \mathbb{R}, G(u) \equiv 0, \eta = 0$.)

Exercise. In the Linearization Theorem 2.6.1, show that *minimum* can be weakened to *quasimin*. (Note that

$$f(a + \alpha d + o(\alpha)) - f(a) = f'(a)\alpha d + o(\alpha) < o(\alpha)$$

since $f'(a)d < 0$.)

5.1.1 Theorem. In problem (#), if $G(\eta) \in K$ and (KT) holds at η, namely

$$H'(\eta) = \pi G'(\eta), \quad \pi G(\eta) = 0, \quad \pi \in K^*,$$

then (#) has a (QM) at η. Conversely, (QM) implies (KT), under the additional assumptions that $G(u) \in K$ is locally solvable at η and that $G'(\eta)(K^*)$ is weak * closed.

Proof. Assume (KT). Suppose that η is not a (QM) of (#). Then there is some sequence $\{z_n\}$ with $\{\|z_n\|\} \to 0$, $G(\eta + z_n) \in K$, and whenever $\theta(z) = o(\|z\|), H(\eta + z_n) - H(\eta) - \theta(z_n) < 0$. From the Fréchet differentiability,

$$\pi G'(\eta)z_n = \pi[G(\eta + z_n) - G(\eta) + \rho(z_n)] \geqslant 0 + \pi\rho(z_n)$$

using $\pi G(\eta) = 0, \pi \in K^*, G(\eta + z_n) \in K$; and also, using $H'(\eta) = \pi G'(\eta)$,

$$H(\eta + z_n) - H(\eta) = H'(\eta)z_n + \sigma(z_n)$$
$$= \pi G'(\eta)z_n + \sigma(z_n) \geqslant \pi\rho(z_n) + \sigma(z_n)$$

as $n \to \infty$. This contradicts $H(\eta + z_n) - H(\eta) - \theta(z_n) < 0$ as $n \to \infty$, when $\theta(z) = \pi\rho(z) + \sigma(z)$. Hence η is a (QM).

Conversely, assume (QM) and the additional hypotheses. Let $G(\mu) + G'(\eta)d \in K$. By local solvability, there is an arc $u = \eta + \alpha d + o(\alpha)$ $(\alpha \geqslant 0)$ with $G(u) \in K$, so (QM) shows that

$$H(\eta + \alpha d + o(\alpha)) - H(\eta) \geqslant o(\|\alpha d + o(\alpha)\|) = o(\alpha),$$

so $H'(\eta)d \geqslant 0$. Thus $G(\eta) + G'(\eta)d \in K$ implies $H'(\eta)d \geqslant 0$. Then (KT) follows from Farkas's theorem (2.2.6), since $G'(\eta)(K^*)$ is weak $*$ closed.

5.1.2 Remarks. The converse holds (see Appendix A.6) when Fréchet differentiability for H is weakened to Hadamard differentiability; this is useful in some control problems. The additional hypotheses for the converse are fulfilled, in particular, if $G'(\eta)(U) = Q$. For example, for the constraint $1 - |u(t)|^2 \in \mathbb{R}_+$ ($\forall t \in I$), $G'(u)z(t) = - 2u(t)z(t)$; hence $G'(\eta)(U) = Q$ iff $\eta(t)$ never vanishes.

Exercise. Discuss this for vector valued $u(t)$.

5.2 Abstract Hamiltonian theory

Consider, as in 1.10, an abstract control problem:

(OC): $\quad\displaystyle\operatorname*{Minimize}_{x \in X, u \in U} \{F(x, u) : Dx = M(x, u),$

$$G(u) \in K, N(x) \in J\},$$

where X, U, W, Q, P are normed spaces (with W, Q, P complete), K and J are convex cones; $D : X \to W$ is a continuous linear map, and $F : X \times U \to \mathbb{R}$, $M : X \times U \to W$, $G : U \to Q$, and $N : X \to P$ are Fréchet differentiable. Denote by $\tilde{X}$ the completion of X. Let $\xi \in X$ and $\eta \in U$. Assume that

(i) the linear map $T = [D - M_x(\xi, \eta) \quad - M_u(\xi, \eta)]$ maps $\tilde{X} \times U$ *onto* a closed subspace of W; and

(ii)   the cones K and J have nonempty interiors.

Then (see 4.4.1) a necessary condition for (OC) to attain a minimum at $(x, u) = (\xi, \eta)$ is (FJ), namely that there exist Lagrange multipliers $\tau \in \mathbb{R}_+$, $\bar{\lambda} \in W'$, $\bar{\mu} \in K^*$, $\bar{\nu} \in J^*$, not all zero, satisfying

(OCF): $\begin{cases} \tau F_x - \bar{\lambda}(D - M_x) - \bar{\nu} N_x = 0; & \bar{\nu} N(\xi) = 0; \\ \tau F_u - \bar{\lambda}(-M_u) + \bar{\mu}(-G_u) = 0; & \bar{\mu} G(\eta) = 0; \end{cases}$

where the partial derivatives F_x, M_u, etc. are evaluated at (ξ, η).

The *adjoint* of the continuous linear map $D : X \to W$ is the continuous linear map $D^T : W' \to X'$, defined by $(D^T w')x = w'(Dx)$ for all $x \in X$ and all $w' \in W'$. (If, for example, X is a space of differentiable functions on the real interval I, which vanish at the ends of I, and if $D = \mathrm{d}/\mathrm{d}t$, then integration by parts shows that $D^T = -\mathrm{d}/\mathrm{d}t$.) The first equation of (OCF) then gives

$$D^T\bar{\lambda} - [\tau F_x + \bar{\lambda}M_x] = -\bar{\nu}N_x.$$

Define the *abstract Hamiltonian* $H(u) = \tau F(\xi, u) + \bar{\lambda}M(\xi, u)$. Set

$$p = [\tau \mathrel{\vdots} \bar{\lambda}], \text{ and } B(x, u) = \begin{bmatrix} F(x, u) \\ M(x, u) \end{bmatrix};$$

then $H(u) = pB(\xi, u)$. Associate also to (OC) the associated problem

(OC#):   Minimize $\{H(u) : G(u) \in K\}$.
          $u \in U$

Then the first two parts of (OCF) are equivalent to the *adjoint differential equation*:

$$D^T\bar{\lambda} - pB_x(\xi, \eta) = -\bar{\nu}N_x(\xi); \quad \bar{\nu}N(\xi) = 0; \quad \bar{\nu} \in J^*.$$

The last two parts of (OCF) are equivalent to $H'(\eta) = \bar{\mu}G'(\eta)$, $\bar{\mu}G(\eta) = 0$, and thus to (KT) for (OC#) at η.

The condition $\bar{\nu}N(\xi) = 0$ for the adjoint differential equation means that the term $-\bar{\nu}N_x(\xi)$ makes a nonzero contribution only at points ξ where $N(\xi)$ is on the boundary of J. If D is a linear differential operator on a space X of functions, denote by D_0 the differential operator, acting on real functions, given by the same formal expression (such as d/dt) as D; let x_0 be a solution of $D_0 x_0 = F(x, u)$ and $\bar{x}_0$ of $D_0 \bar{x}_0 = F(\xi, \eta)$. Define also

$$y = \begin{bmatrix} x_0 \\ x \end{bmatrix}, \quad \tilde{B}(y, u) = B(x, u), \quad Dy = \begin{bmatrix} D_0 x_0 \\ Dx \end{bmatrix}, \quad \zeta = \begin{bmatrix} \bar{x}_0 \\ \xi \end{bmatrix}.$$

Now

$$p\tilde{B}_y = [0 \mathbin{\vdots} \tau F_x + \bar{\lambda} M_x] \quad \text{and} \quad D^T p = [0 \mathbin{\vdots} D^T \bar{\lambda}],$$

since $\tau \in \mathbb{R}_+$ is a constant. Hence the adjoint differential equation may also be written as:

$$D^T p = p\tilde{B}_y(\zeta, \eta) - \bar{\nu} N_y(\xi); \quad \bar{\nu} N(\xi) = 0; \quad \bar{\nu} \in J^*.$$

The relations between solutions of (OC), (OCF), and (OC#), under the hypotheses stated above, may be summarized as follows.

(OC) has a minimum (ξ, η)

(A) $\Downarrow$ $\Uparrow$ (B)

conditions   (OCF)

$\Downarrow$ $\Uparrow$

adjoint differential equation, and (KT) for (OC#) at η

(D) $\Downarrow$ $\Uparrow$ (C)

adjoint differential equation, and (OC#) has quasimin at η

(E) $\Downarrow$ $\Uparrow$

adjoint differential equation, and (OC#) has minimum at η.

The implications shown assume the following hypotheses. (A) assumes hypotheses (i) and (ii), stated at the beginning of 5.2. (If, instead, int $K = \emptyset$, then $G(u) \in K$ may be assumed locally solvable.) (D) follows from Theorem 5.1.1, with no additional hypothesis. (C) assumes that $G(u) \in K$ is locally solvable at η, and $G'(\eta)(K^*)$ weak $*$ closed, using 5.1.1. For (E), it suffices if the problem (OC#) is convex; for (B) theorem 4.5.2 may be applied, if suitable convexity is assumed. Alternative sufficient hypotheses for (E) and (B), not involving any convexity, are obtained in section 5.3.

Therefore, under the indicated hypotheses (A) and (E), a minimum of (OC) implies a minimum of (OC#), together with the adjoint differential equation. This is *Pontryagin's principle* (as a *necessary* condition), for the problem (OC). The necessary condition becomes also *sufficient* under additional hypotheses ((C) and (B)).

It is often useful to use the L^1-norm for a space of functions on an interval (or compact subset of $\mathbb{R}^r$), I; this norm is

$$\|s\|_1 = \int_I \|s(t)\| \mathrm{d}t.$$

But the convex cone of nonnegative functions in the space $L^1(I)$ has empty interior. A hypothesis such as int $K \neq \emptyset$ can, however, sometimes be weakened (see Appendix A.4); also weaker derivatives than Fréchet can sometimes be used (see Appendix A.6).

5.3 Pointwise theorems

Assume now, as in 1.10, that X and U are spaces of functions on a real interval $I = [0, T]$, with F and M defined from functions f and m. Then often (see 1.10 and 1.8) the functional $\bar{\lambda}$ can be represented by a function $\lambda(\cdot)$ on I, and then the *abstract Hamiltonian* H can be written as an integral

$$H(u) = \int_I h(u(t), t) \, \mathrm{d}t,$$

where the *Hamiltonian h* is

$$h(u(t), t) = f(\xi(t), u(t), t) - \lambda(t)m(\xi(t), u(t), t) \ (t \in I).$$

Assume also that $G(u) \in K$ iff $(\forall t \in I) \, g(u(t), t) \in S$.

It is now shown that, if the space U has the $L^1(I)$ norm, $\|u\|_1 = \int_I \|u(t)\| \, \mathrm{d}t$, and if U has the *interpolable* property, that if $u \in U$ and $v \in U$, then also $w \in U$, where $w(t) = u(t)$ for $t \in A \subset I$ and $w(t) = v(t)$ for $t \in I \backslash A$, then (QM) for (OC#) implies a minimum of $h(u(t), t)$ *almost everywhere* (i.e. for all $t \in I$ except for a set of zero measure; the measure of a set $A \subset I$ is $\int_A dt$.) Under these assumptions, (QM) for (OC#) implies a *minimum* for (OC#). The following theorem 5.3.1 holds also when I is, instead, a compact subset of $\mathbb{R}^n$, and D is a linear partial differential operator.

5.3.1 Theorem. Let U have the $L^1(I)$-norm, and be interpolable; let H, G, K be as specified above; let h satisfy a Lipschitz condition with respect to u. If (OC#) has a (QM) at $u = \eta$, then $h(u(t), t)$ is minimized, almost everywhere in I, with respect to $u \in G^{-1}(K)$, at $u = \eta$. Consequently (OC#) attains a minimum at $u = \eta$.

Proof. Suppose not; then for some $\bar{u} \in G^{-1}(K)$, $h(\bar{u}(t), t) < h(\eta(t), t)$ for all $t \in A^{\#}$, a set of positive measure. From the Lipschitz hypothesis

$$(\forall t \in I) - h(\bar{u}(t), t) + h(\eta(t), t) = \phi(t)\|\bar{u}(t) - \eta(t)\|,$$

where ϕ is a bounded function, and $\phi(t) = 0$ where $\bar{u}(t) = \eta(t)$. If ϕ is continuous at $t_0 \in A^{\#}$, then there are $A \subset A^{\#}$ and $\sigma > 0$ for which $\phi(t) \geq \sigma$ for $t \in A$, and A has positive measure. Define the continuous arc u_β ($\beta \geq 0$) by

$$u_\beta(t) = \begin{cases} \bar{u}(t) & \text{for } t \in A_\beta \equiv \{t \in A : |t - t_0| \leq \psi(\beta)\} \\ \eta(t) & \text{otherwise,} \end{cases}$$

choosing $\psi(\beta)$ so that, as $\beta \downarrow 0$, the $L^1(I)$-norm $\|u - \eta\|_1 = \beta$. Then $u_0 = \eta$, $u_\beta \in U$ for $\beta \geq 0$ since U is interpolable, and $G(u_\beta) \in K$. Then

$$\begin{aligned} H(\eta) - H(u_\beta) &\geq \int_{A_\beta} \phi(t)\|\eta(t) - \bar{u}(t)\| \, dt \\ &\geq \sigma \int_{A_\beta} \|\eta(t) - \bar{u}(t)\| \, dt \\ &= \sigma\|u_\beta - \eta\|_1 = \sigma\beta. \end{aligned}$$

But this contradicts the requirement of a (QM), that

$$H(u_\beta) \geq H(\eta) + o(\beta) \quad \text{as } \beta \downarrow 0.$$

Appendix A.5 shows that, when ϕ is not continuous, appropriate sets A_β can be constructed so that the remainder of this proof is valid. From a minimum of h, almost everywhere, a minimum of H immediately follows.

5.3.2 Remarks. The Lipschitz condition holds if h has a continuous partial derivative with respect to u.

Assumption (E) may be that U has the $L^1(I)$-norm, where I is compact and the measure is Lebesgue, U is interpolable, and h satisfies the Lipschitz condition.

5.3.3 Remarks. Suppose that (OCF) holds, and also that $\tau = 1$; then, by theorem 5.1.1, (OC) has a (QM) at (ξ, η). Suppose also that the differential equation $Dx = M(x, u)$ is *uniquely solvable*, meaning that for each $u \in U$ with $G(u) \in K$ and $\|u - \eta\|$ sufficiently small, $Dx = M(x, u)$ (which includes the boundary conditions on x) has a solution $x = \Phi(u)$ say, which is unique. Then

$$F(x, u) = \int_I f(\Phi(u)(t), u(t), t)\, dt$$

has the form of H in theorem 5.3.1, noting that f is continuously differentiable; also $N(x) \in J$ iff $n \circ \Phi(u(t), t) \in V$. Then Theorem 5.3.1 shows that (OC) has a minimum at (ξ, η).

Hence hypothesis (B) in 5.2 may be *either* that (OC) satisfies, at (ξ, η), the convexity conditions of Theorem 4.5.3, *or* that $\tau = 1$; f, g, m, h are given by the integral expressions discussed (as in 1.10); U has the $L^1(I)$-norm and is interpolable; and $Dx = M(x, u)$ is uniquely solvable.

5.3.4 Theorem. Under hypotheses (A) and (E), necessary conditions for a minimum of (OC) are the adjoint differential equation, and a minimum of (OC#). These conditions are also sufficient under the additional hypotheses (B) and (C).

Proof. As above. (Refer also to 5.2.)

Exercise. Set out in detail the first alternative for hypothesis (B).

Exercise. Consider the version of (OC) in 1.10, with f taken as a convex quadratic function, and m, g, n as linear functions; assume hypotheses (i) and (ii) of 5.2. Verify that the necessary and sufficient conditions of Theorem 5.3.4 hold.

Exercise. If U is the subspace of $L^1(I)$ consisting of the *bounded* (measurable) functions on I, show that U is interpolable.

5.3.5 Remark. The hypothesis that $T(\tilde{X} \times U) = W$, which implies (i) in 5.2, implies that the *linear* differential equation

$$[D - M_x(\xi, \eta)]x - M_u(\xi, \eta)u = w$$

has, for each $w \in W$, a solution $x \in \tilde{X}$, for some $u \in U$; note that this x must satisfy the boundary conditions included in the specification of X. (See also 5.3.9.)

5.3.6 Example. Suppose that F and M are each linear in u, and that $A = G^{-1}(K)$ is convex and compact. Then (OC#) minimizes a continuous linear function $H(u)$ over a compact set, so the minimum is reached, say at $u = \eta$. The Krein–Milman theorem states that, for A compact convex, $A = \overline{\mathrm{co}}\, E$ where $E = \mathrm{extr}\, A$. A calculation similar to 2.1.4 then shows that $H(\eta) \geqslant \inf_{e \in E} H(e) = c$. But c is attained at some point of $\bar{E}$, since $\bar{E}$ is compact; and $H(\eta) \leqslant c$ by definition of η. Hence (OC#) attains its minimum at a point of $\bar{E}$. Now assume also hypothesis (A) of Theorem 5.3.4. If a unique optimal control η exists, then η must uniquely minimize the convex problem (OC#): hence $\eta \in \bar{E}$. In particular, if U consists of piecewise continuous real functions, and the control region is given by $|u(t)| \leqslant 1$, then E, and hence $\bar{E}$, consists of functions whose only values are ± 1. So the optimal control in this case, if it exists uniquely, is *bang-bang*.

Exercise. Discuss the optimal inventory model of 1.5, using Pontryagin methods (i.e. Theorem 5.3.4),. Here $\{s_t\}$ may be taken as the trajectory x, and $\{m_t\}$ as the control function u.

5.3.7 Remark. The formal analysis of an optimal control problem, given in 1.10, is validated, for uniform norms, by Theorem 4.4.1, assuming hypotheses (i) and (ii) of 5.2, and noting that the cone K has nonempty interior if $\text{int } S \neq \emptyset$, and similarly for V and J.

5.3.8 Remark. When (OC) is considered with $L^1(I)$-norms, the constraint $(\forall t \in I)\, g(u(t), t) \in S$ becomes $G(u) \in K$, where $G(u)(t) = g(u(t), t)$, and G is now linearly Gâteaux differentiable, but not necessarily Fréchet differentiable. Appendix A.6 shows that, if g satisfies the Lipschitz condition

$$(\forall z_1, z_2)\ \|g(z_1, t) - g(z_2, t)\| \leq k\|z_1 - z_2\|,$$

where the constant k does not depend on z_1 and z_2, and if also $\int_I |g(0, t)|\, dt$ is finite, then the (FJ) result of 4.4.1 remains valid with the linear Gâteaux derivative in place of a Fréchet derivative. A similar comment applies to F.

Also $\text{int } S \neq \emptyset$ does not, with an L^1-norm, imply that $\text{int } K \neq \emptyset$. For example, if $S = \mathbb{R}^k_+$, then K is the cone of non-negative functions in $L^1(I)$, and $\text{int } K = \emptyset$.

Exercise. For (OC) as in 1.10, with $D = d/dt$, use integration by parts, and the boundary conditions on X, to show that $D^T = -D$. (Consider

$$(D^T s)(q) = s(Dq) = \int_0^T s'(t)q(t)\, dt = -\int_0^T s(t)q'(t)\, dt.$$

Deduce that the adjoint differential equation leads to

$$\frac{d\lambda(t)}{dt} = \tau f_x(\xi(t), \eta(t), t) - \lambda(t)m_x(\xi(t), \eta(t), t)$$

$$- \nu(t)n_x(\xi(t), t),\ \nu(t) \in V^*,\ \nu(t)n(\xi(t), t) = 0\ \ (0 \leq t \leq T).$$

In particular, if $V = \mathbb{R}^h_+$, then $\nu(t) \geq 0$; and the term $\nu(t)n_x(\xi(t), t)$ is zero except when $n(\xi(t), t) = 0$.

Exercise. If the Pontryagin theory is applied to the linear program

Minimize $\{c^T x + d^T u : Dx = Ax + Bu + b : x \geqslant 0, |u| \leqslant 1\}$,
$\quad$ x, u

where D, A, B are matrices, and $|u| \leqslant 1$ applies to each component of u, show that there results an 'adjoint equation'

$$-(D^T \lambda)^T = \tau c^T - \lambda^T A - v^T = 0, v \geqslant 0, v^T x = 0;$$

and the associated minimization problem

$$\text{Minimize } \{(\tau d - B^T \lambda)^T u : |u| \leqslant 1\}.$$
$\quad$ u

In this *linear* problem, $\tau = 1$ can be assumed. It then follows that u_j has the sign of $(d - B^T \lambda)_j$. How does this formulation relate to the dual of the given linear program?

5.3.9 Example. The differential equation

$$\frac{dz}{dt} + (1 - u(t)^2)z(t) = 0, z(0) = 0, z(1) = c > 1,$$

has a solution for $c \geqslant 1/e$, but not otherwise; $c = 1/e$ is reached with $u(t) \equiv 0$. For given $c \geqslant 1/e$, let $x = z - y$ where y is a linear function such that $x(0) = x(1) = 0$; then, for suitable ψ,

$$\frac{dx(t)}{dt} = m(x(t), u(t), t) \equiv -(1 - u(t)^2)x(t) - \psi(t).$$

For this equation to be locally solvable at $(x, u) = (\xi, \eta)$, it suffices that

$$\frac{dx(t)}{dt} - [-(1 - \eta(t)^2)x(t) + 2\eta(t)\xi(t)u(t)] = w(t)$$

be solvable for each $w \in C[0, 1]$, with $x(0) = x(1) = 0$. This does not happen for the 'boundary' case $c = 1/e, \eta(t) \equiv 0$; and the original system is clearly not locally solvable in this case. Local solvability holds when $c > 1/e, \eta(t) > 0$.

5.4 Problems with variable endpoint

The abstract theory of 5.2 applies not only to optimal control problems such as in 1.10, where the differential equation relates to a fixed time interval $[0, T]$, or more generally to a fixed region in a space of higher dimension, but also to various problems where the endpoint is variable. Consider a differential equation

$$\frac{dx(t)}{dt} = m(x(t), u(t), t) \quad (t \in \mathbb{R}_+),$$

with the initial condition $x(0) = 0$ on the trajectory. Define a *terminal region* by $q(x(t)) \leqslant 0$, where q is a suitable real function. Thus the fixed interval $[0, T]$ is now replaced by a variable interval $[0, t_F]$, where $q(x(t_F)) = 0$ and $q(x(t)) > 0$ for $0 < t < t_F$. Denote by T_F the time t_F when the optimal trajectory $x(t) = \xi(t)$ first reaches the terminal region. Let $\pi(s) = 1$ if $s \geqslant 0$, and let $\pi(s) = 0$ if $s < 0$. If it is further assumed that $q(\xi(t)) < 0$ for some interval $(T_F, T_F + \delta)$, then the objective function becomes

$$F(x, u) \equiv \int_0^{t_F} f(x(t), u(t), t) \, dt$$

$$\equiv \int_0^{T} f(x(t), u(t), t) \, \pi \circ q(x(t)) \, dt$$

for $\|x - \xi\|$ sufficiently small, where $T > T_F$ is a suitable constant. In particular, when $f(\cdot) \equiv 1$, the objective function to be minimized is t_F, the time for the trajectory to first reach the terminal region, subject to the constraints of the problem. The constraints will be taken as the above differential equation, with initial condition $x(0) = 0$, and (as in 1.10)

$$g(u(t), t) \in S \quad \text{and} \quad n(x(t), t) \in V \quad (0 \leqslant t \leqslant t_F).$$

Equip the space of trajectories with the norm $\|x\| = \|x\|_\infty + \|Dx\|_\infty$, where $D = d/dt$ (as in 1.10). On the assumption that $(q \circ \xi)'(T_F) \neq 0$, T_F is an isolated zero of $q \circ \xi$; Appendix A.3 shows then that, for sufficiently small $\|x - \xi\|$, $q \circ x$ has a zero t_F, where $t_F(x)$ is a *Fréchet differentiable* function of $x \in X$ at $x = \xi$, and $t_F(\xi) = T_F$. The assumption implies that, for sufficiently small $\|x - \xi\|$, $q(x(t)) < 0$ for some interval $(t_F, t_F + \delta(x))$; this excludes some cases where ξ is tangential to the terminal region. Now if $t_F(\cdot)$ is Fréchet differentiable, and if $f(\cdot, \cdot, \cdot)$ is continuously Fréchet differentiable, then it follows readily that $F(x, u)$ is Fréchet differentiable with respect to (x, u). Therefore the theory of 5.2 and 5.3 can be applied to this variable-endpoint problem.

For $x = \xi$, and $0 \leqslant t < T_F$, $\pi \circ q(\xi(t)) = 1$, and hence the Hamiltonian equals

$$h(u(t), t) = \tau f(\xi(t), u(t), t)(\pi \circ q)(\xi(t)) - \lambda(t)m(\xi(t), u(t), t)$$
$$= \tau f(\xi(t), u(t), t) - \lambda(t)m(\xi(t), u(t), t);$$

and the adjoint differential equation becomes

$$\frac{d\lambda(t)}{dt} = \tau f_x(\xi(t), \eta(t), t) - \lambda(t)m_x(\xi(t), \eta(t), t)$$

$$- \nu(t)n_x(\xi(t), t), \nu(t) \in V^*, \nu(t)n(\xi(t), t) = 0 \quad (0 \leqslant t \leqslant T_F).$$

If $F(x, u) = t_F(x)$, then $f(\cdot) \equiv 1$, so

$$h(u(t), t) = - \lambda(t)m(\xi(t), u(t), t),$$

and the adjoint equation omits the term in f_x.

The adjoint of d/dt is $- d/dt$, by integrating by parts, if the boundary conditions are $x(0) = 0 = x(T_F)$. To attain the latter, consider a change of variable from x to $\tilde{x} = x - y$, with y such that $\tilde{x}(0) = \tilde{x}(T_F) = 0$. Then the theory can be applied with trajectory $\tilde{x}$. Changing the variable back to x recovers the results in the last paragraph.

Exercise. Consider $F = t_F(x)$ with the linear differential equation

$$\frac{dx(t)}{dt} = a(t)x(t) + b(t)u(t).$$

Then $h = \tau - (a\xi + bu)$, and the adjoint equation is

$$\frac{d\lambda(t)}{dt} = - \lambda(t)a(t) - \nu(t)n_x(\xi(t), t).$$

Discuss this case, in relation to bang-bang control, when n also is linear.

Exercise. In the last exercise, replace F by

$$F(x, u) = \int_0^{t_F} \Theta(x(t))\, dt,$$

for a suitable function Θ.

Exercise. Suppose instead that the terminal region is a closed polyhedron in $\mathbb{R}^n$, specified by inequalities $q_i(x) \leq 0$ ($i = 1, 2, \ldots, k$). Then, for some i_0, $q_{i_0}(\xi(T_F)) = 0$ while $q_i(\xi(T_F)) \geq 0$ for each $i \neq i_0$. Discuss the application of the above theory with q_{i_0} in place of q.

Exercise. Consider an optimal control problem with a *partial* differential equation, thus with D a linear *partial* differential operator, and a terminal region specified by $q(x(\cdot)) \in -A$, where A is a closed convex cone. When can the above theory be generalized to this case? (Use the 'extension' in Appendix A.3.)

Exercise. What happens when the initial condition $x(0) = 0$ on the trajectory x be replaced by $\psi(x(0)) = 0$; thus the trajectory begins on a curve $\psi(\cdot) = 0$.

References

Berkovitz, L.D. (1974), *Optimal Control Theory*, Springer, New York.
Craven, B.D. (1977), Lagrangean conditions and quasiduality, *Bull. Austral. Math. Soc.*, **16**, 325–339.
Luenberger, D.G. (1969), *Optimization by Vector Space Methods*, Wiley, New York.

Notes

See Craven (1977) for some of the underlying theory. Refer to Berkovitz (1974) for other approaches to the Pontryagin theory, and also theorems on existence of optimal solutions. Luenberger (1969) gives in Section 9.6 another approach to the Pontryagin principle.

CHAPTER SIX

Fractional and complex programming

6.1 Fractional programming

A number of applications lead to constrained minimization problems, in which the objective function, to be minimized or maximized, is a quotient, $f(x)/g(x)$, of two functions. Such a problem is called a *fractional programming problem*. In particular, it is a *linear fractional programming problem* if f and g are linear, or affine, functions, and the constraints are linear. Although such problems are particular cases of non-linear programming problems, stronger results for fractional programming problems are obtainable by proceeding directly, rather than applying the theory of Chapter 4. This applies both to theoretical questions, notably duality theory, and to effective algorithms for computing an optimum.

Some applications are outlined as follows. Consider the *cutting stock problem* (Gilmore and Gomory, 1961, 1963; see references at end of chapter) of cutting rolls of paper (or billets of steel, etc.) into smaller rolls (or sheets, or billets, etc.) of specified sizes and numbers, in such a way as to minimize wastage. This leads to a linear programming problem of the form

$$\text{Minimize } c^T x \text{ subject to } Ax = b, x \geqslant 0,$$

where each column of A represents a possible way of cutting a roll into smaller rolls. If, instead, the ratio of wastage to useful output is to be minimized, then the objective function $c^T x$ becomes instead a fraction $c^T x / d^T x$, leading to a linear fractional program. (The actual problem is made more complicated by the need to generate columns of A, when required, by a suitable algorithm, and often also by the requirement for integer solutions x; these aspects are not discussed here.)

A similar problem arises in optimizing shipping schedules (Pollack, Novaes, and Frankel, 1965; Bitran and Novaes, 1973). A linear programming model may be set up to find the best combination of cargoes to be loaded into a ship, in terms of maximum profit. This leads to a linear programming model, which may however not be realistic, since it disregards the fact that some kinds of cargo are loaded, or unloaded, more slowly than others. To allow for this, the ratio of profit per trip to total trip time may be maximized, subject to appropriate linear constraints. This leads to a linear fractional program, with an objective function $(c^T x + \alpha)/(d^T x + \beta)$.

The problem of finding optimal (cost minimizing) policies for the management of stochastic systems, describable as Markov chains, has been studied by Derman (1962), Klein (1963), and Fox (1966). These also lead to linear fractional programs.

Bereanu (1963) has considered a stochastic programming problem, arising in an agricultural application, where the objective function to be maximized is the probability that a certain set of linear inequalities is satisfied. This leads to a nonlinear fractional programming problem with objective function of the form $(c^T x + \alpha)/(x^T x)^{1/2}$ (in the present notation). If $\alpha = 0$, then both numerator and denominator are homogeneous functions of degree 1.

Ziemba, Parkan, and Brooks-Hill (1974) have discussed the problem of optimum portfolio selection for an investor, under certain economic assumptions. The authors use results of Lintner (1965) to reduce their economic optimization model to a combination of a nonlinear fractional programming problem with linear constraints and an objective of the form

$$(c^T x + \alpha)(x^T A x)^{1/2},$$

and a stochastic optimization problem involving only one random variable. The matrix A is positive definite; the denominator arises from the standard deviation of a random variable.

Meister and Oettli (1967) have applied fractional programming to calculating the maximum rate of information transmission through an information channel. This leads to a problem with linear constraints, and an objective of the form

$$(c^T x - t \log t)/(d^T x)$$

(in the present notation). Isbell (1956) has applied fractional programming to game theory. Schaible (1976) gives an extensive bibliography of fractional programming, theory and applications.

The next three sections deal with theory and algorithms for fractional programming problems. (The economic or statistical details, necessarily cited in the above outline of applications, will not be discussed further.)

6.2 Linear fractional programming

Consider the linear fractional programming problem:

(LFP): Maximize $\dfrac{c^T x + \alpha}{d^T x + \beta}$ subject to $x \geqslant 0, Ax \leqslant b,$

and $d^T x + \beta > 0.$

Here $x, c, d \in \mathbb{R}^n; A \in \mathbb{R}^{m \times n}; b \in \mathbb{R}^m; \alpha, \beta \in \mathbb{R}$. To agree with most publications on the theory of fractional programming, the problem has been expressed as a maximization problem; the 'linear' functions such as $c^T x + \alpha$ are actually affine. If values of x with $d^T x + \beta = 0$ were allowed, then the objective function would either take infinite values, or be indeterminate (thus 0/0); the constraint $d^T x + \beta > 0$ is included to exclude these cases. Here $Ax \leqslant b$ means

$b - Ax \in \mathbb{R}^m_+$; this could easily be extended to more general cones.

The two mathematical programming problems

(I):   Maximize $f(x)$ subject to $x \in S$

(II):   Maximize $F(\xi)$ subject to $\xi \in T$

will be called *q-equivalent* if there is a one-one map q of the constraint set S of (I) onto the constraint set T of (II), such that $f(x) = F(q(x))$ for each $x \in S$. Clearly, q-equivalence defines a reflexive transitive relation on the set of maximizing problems. A similar definition applies to minimizing problems.

Exercise. For the linear fractional program

$$\text{Maximize } (c^T x + \alpha)/(d^T x + \beta) \text{ subject to } Ax \leqslant b,$$

when does the transformation $x = k + My$ lead to a q-equivalence? When is the given problem thus q-equivalent to one lacking constant terms α, β?

6.2.1 Lemma. Let (I) and (II) be q-equivalent. Then (a) if (I) attains a maximum at x^*, then (II) attains a maximum at $q(x^*)$; and (b) if (II) attains a maximum at ξ^*, then (I) attains a maximum at $q^{-1}(\xi^*)$.

Proof. For (a), let $\xi = q(x)$ and $\xi^* = q(x^*)$; for (b), let $x = q^{-1}(\xi)$ and $x^* = q^{-1}(\xi^*)$. Since q maps S one-one onto T, these definitions are equivalent, and

$$[x \in S \ \& \ x^* \in S \ \& \ f(x) \leqslant f(x^*)]$$
$$\Leftrightarrow [\xi \in T \ \& \ \xi^* \in T \ \& \ F(\xi) \leqslant F(\xi^*)].$$

6.2.2 Remark. This lemma applies to global maxima. It applies also to local maxima, if q and q^{-1} are continuous, for then q maps a sufficiently small neighbourhood of x^* onto a neighbourhood of ξ^*.

6.2.3 Theorem. The linear fractional program (LFP) is q-equivalent to the linear program

(LFEP): Maximize $c^T y + \alpha t$ subject to $y \geqslant 0, t \geqslant 0$,

$$d^T y + \beta t = 1, Ay - bt \leqslant 0;$$

where $y \in \mathbb{R}^n_+$, $t \in \mathbb{R}_+$, and q is defined by $q(x) = (y, t)$ with

$$t = (d^T x + \beta)^{-1}, \quad y = xt;$$

provided that, for each $y \geqslant 0$, the point $(y, 0)$ is not feasible for (LFEP).

Proof. (see Charnes and Cooper, 1962; Craven and Mond, 1975). If x is feasible for (LFP) then $(y, t) = q(x)$ satisfies

$$y \geqslant 0, t > 0, Ay - bt = t(Ax - b) \leqslant 0, d^T y + \beta t = 1;$$

so (y, t) is feasible for (LFEP). If (y, t) is feasible for (LFEP) then, since $t > 0$ by the hypothesis that $(y, 0)$ is not feasible, $x = q^{-1}(y, t)$ is defined by $x = y/t$, with x finite, and then

$$x \geqslant 0, Ax - b = t^{-1}(Ay - bt) \leqslant 0, d^T x + \beta = t^{-1} > 0;$$

so x is feasible for (LFP). Since also

$$(c^T x + \alpha)/(d^T x + \beta) = (c^T y + \alpha t)/(d^T y + \beta t) = (c^T y + \alpha t)/1,$$

(LFP) and (LFEP) are q-equivalent.

6.2.4 Remarks. The optimum of (LFP) can be found by optimizing one linear program. The hypothesis that $(y, 0)$ is not feasible for (LFEP) holds if (LFP) has a *bounded* non-empty constraint set S. For if $\bar{y} \geqslant 0$ and $(\bar{y}, 0)$ is feasible for (LFEP) then, for each $x \in S$, $x + \lambda \bar{y} \in S$ for each $\lambda > 0$ since $A\bar{y} \leqslant 0, \bar{y} \geqslant 0$; and this contradicts the boundedness of S.

Consider now the special case of (LFP):

(SLFP): Maximize $\dfrac{c^T x}{d^T x}$ subject to $x \geqslant 0, Ax \leqslant b$,

and $d^T x > 0$.

Example. (LFP) may be expressed in the form of (SLFP) by writing

$$\frac{c^T x + \alpha}{d^T x + \beta} = \frac{c^T x + \alpha s}{d^T x + \beta s},$$

adjoining the additional constraints $s \leqslant 1$ and $-s \leqslant -1$ ($s \in \mathbb{R}$).

Associate to (SLFP) the linear fractional program:

(DLFP):   Minimize $\dfrac{c^T u}{d^T u}$ subject to $u \geqslant 0, v \geqslant 0$,

$$b^T v \leqslant 0, A^T v - (cd^T - dc^T)u \geqslant 0, d^T u > 0.$$

6.2.5 Theorem (Duality). Let $x \geqslant 0$ and $Ax \leqslant b$ imply $d^T x > 0$; let (SLFP) attain its (finite) maximum at $x = x^*$; then (DLFP) is a dual problem to (SLFP).

Proof. From the constraints of the two problems,

$$0 \geqslant b^T v \geqslant v^T Ax = x^T A^T v \geqslant (c^T x)(d^T u) - (d^T x)(c^T u)$$

whenever x is feasible for (SLFP) and (u, v) is feasible for (DLFP). Hence

$$\frac{c^T x}{d^T x} \leqslant \frac{c^T u}{d^T u};$$

thus weak duality holds for (SLFP) and (DLFP).

Let $x = x^*$ optimize (SLFP): define the pair of dual linear programs:

(i):   Maximize $F(x) = (d^T x^* c - c^T x^* d)^T x$

            subject to $x \geqslant 0, Ax \leqslant b$;

(ii):   Minimize $G(v) = b^T v$ subject to

            $v \geqslant 0, A^T v \geqslant d^T x^* c - c^T x^* d.$

If $Ax \leqslant b, x \geqslant 0$, and $F(x) > F(x^*)$, then $d^T x > 0$, and

$$\frac{c^T x}{d^T x} > \frac{c^T x^*}{d^T x^*},$$

which is contradicted, since (SLFP) attains its maximum at $x = x^*$. Hence (i) is maximized at $x = x^*$. By the duality theorem for linear programming (3.2.3), $F(x^*) = b^T v^*$ for some v^* feasible for (ii); hence $b^T v^* = 0$. Hence (u^*, v^*), where $u^* = x^*$, is feasible for (DLFP), and the objective function equals that of (SLFP) at $x = x^*$. This, with weak duality, proves that (DLFP) is a dual to (SLFP).

Exercise. Extend Theorem 6.2.5 from (SLFP) to (LFP).

Exercise. Use 6.2.3 to find a linear program q-equivalent to (DLFP). Does it relate to the linear program q-equivalent to (SLFP)?

6.2.6 Remark. If the constraints $d^T x > 0$, $d^T u > 0$ are weakened to $d^T x \geq 0$, $d^T u \geq 0$, then weak duality still holds, as a limiting case. (Thus if $d^T x = 0$ then $c^T x < 0$, since (SLFP) has a finite maximum; so $c^T x / d^T x = -\infty$.)

6.2.7 Corollary. Let $S = \{x : x \geq 0, Ax \leq b\}$ be bounded; let $x \in S$ imply $d^T x \geq 0$; and let (DLFP) attain a (finite) minimum. Then (SLFP) is a dual problem to (DLFP).

Proof. Since (DLFP) attains a finite minimum, weak duality implies that the values of $c^T x / d^T x$ for $x \in S$ are bounded above; hence $x \in S$ implies S is bounded away from $\{x : d^T x = 0\}$. Since also S is bounded closed, hence compact, (SLFP) attains a finite maximum. The previous theorem then shows that Min(DLFP) = Max(SLFP).

6.2.8 Example.

(SLFP):   Maximize $(x_1 + 2x_2)/(x_1 - x_2)$

      subject to $x_1, x_2 \geq 0$, $x_1 - x_2 \geq 1$, $x_1 + x_2 \leq 2$.

(LFEP): Maximize $y_1 + 2y_2$ subject to $y_1, y_2, t \geq 0$,

$$y_1 - y_2 = 1, y_1 + y_2 \leq 2t, y_1 - y_2 \geq t.$$

The optimal solution (from (LFEP) is $x_1 = 3/2$, $x_2 = 1/2$, Max = 5/2.

(DLFP): Minimize $(u_1 + 2u_2)/(u_1 - u_2)$ subject to

$$u_1, u_2, v_1, v_2 \geqslant 0, -v_1 + v_2 + 3u_2 \geqslant 0, v_1 + v_2 - u_1 \geqslant 0,$$
$$-v_1 + 2v_2 \leqslant 0, u_1 - u_2 > 0.$$

The optimal solution is $u_1 = 3/2, u_2 = 1/2, v_1 = 3/2, v_2 = 0$; Min $= 5/2$. Observe that the constraint $u_1 - u_2 > 0$ necessary in (DLFP), since the other constraints of (DFLP) are satisfied by $v_1 = v_2 = 0, u_1 = 0, u_2 = 1$, for which the objective of (DLFP) is $-2 < 5/2$.

Exercise. Consider (SLFP) from 6.2.8 with the constraint $x_1 + x_2 \leqslant 2$ omitted. Do (i) the hypotheses, and (ii) the conclusion of Theorem 6.2.3 then hold for this instance? Does duality hold?

6.2.9 Remark. There is no *unique* dual to a linear fractional program. The following is a different method of constructing a dual, leading to a different result. The objective function of (LFP) is modified to the form $(c_0^T x - 1)/(d^T x + \beta)$, for suitable c_0, by adding $\lambda = -(1 + \alpha)/\beta$ to it. This modified (LFP) has a q-equivalent linear program, whose dual is

$$\text{Minimize } [0 \vdots -1 \vdots 1][u \vdots v \vdots w]^T \text{ subject to } u, v, w \geqslant 0,$$
$$A^T u + dz \geqslant c_0, -b^T u + \beta z \geqslant -1. \qquad (\#)$$

Now the linear fractional program

$$\text{Maximize } c_1^T z/(d_1^T z + 1) \text{ subject to } z \geqslant 0,$$
$$A_1 z \leqslant b_1, d_1^T z + 1 > 0,$$

has an equivalent linear program

$$\text{Maximize } c_1^T y - t \text{ subject to } y, t \geqslant 0, d_1^T y + t = 1,$$
$$A_1 y - b_1 t \leqslant 0.$$

The latter is q-equivalent, with $q(y, t) = y$, to

$$\text{Maximize } c_1^T y \text{ subject to } y \geqslant 0, \text{ and}$$

$$\begin{bmatrix} -A_1 - b_1 d_1^T \\ -d_1^T \end{bmatrix} y \;\geqslant\; \begin{bmatrix} -b_1 \\ -1 \end{bmatrix} \qquad (\#\#)$$

Now (#) has the form of (##), if c_1, d_1, A_1, b_1 are suitably chosen. If this is done, and then λ subtracted from the objective, a dual of (LFP) is obtained as the linear fractional program:

$$\text{Minimize } \frac{-\lambda b^T u - \alpha z - \lambda}{b^T u - \beta z + 1} \text{ subject to } u \geqslant 0,$$

$$b^T u - \beta z + 1 > 0, (A^T - c_0 b^T)u + (d + c_0 \beta) \geqslant c_0,$$

where $\lambda = -(1 + \alpha)/\beta$ and $c_0 = c + \lambda d$.

Exercise. Calculate this alternative dual for the problem 6.2.8.

6.3 Nonlinear fractional programming

The problem (LFP) generalizes to the nonlinear problem

(NFP): Maximize $f(x)/g(x)$ subject to $h(x) \in S, g(x) > 0$,

where $x \in \mathbb{R}^n$, and $f \colon \mathbb{R}^n \to \mathbb{R}, g \colon \mathbb{R}^n \to \mathbb{R}$, and $h \colon \mathbb{R}^n \to \mathbb{R}^m$ are given functions, and S is a convex cone in $\mathbb{R}^m$. Let $\phi_0 \colon \mathbb{R} \to \mathbb{R}$ be a strictly increasing function, with $\phi_0(t) > 0$ for $t > 0$. Let $T \subset \mathbb{R}^s$ be a convex cone and, for each $t \in \mathbb{R}_+$, let $\psi(t) \colon \mathbb{R}^m \to \mathbb{R}^s$ be a function such that $z \in S$ iff $\psi(t)(z) \in T$. (In particular, when $S = T = \mathbb{R}^m_+$, $\psi(t)$ may be a diagonal matrix with positive diagonal elements for each t.) For $0 \neq t \in \mathbb{R}_+$ and $y \in \mathbb{R}^m$, define

$$F(y, t) = f(y/t)\phi_0(t); \; G(y, t) = g(y/t)\phi_0(t);$$
$$H(y, t) = \psi(t)[h(y/t)].$$

Define $F(y, 0) = \lim_{t \downarrow 0} F(y, t)$, and similarly $G(y, 0)$ and $H(y, 0)$, whenever these limits exist; assume that $G(0, 0) = 0$ whenever it exists. Consider the associated nonlinear programming problem

(NFEP):   Maximize $k^{-1}F(y, t)$ subject to

$$G(y, t) = k, H(y, t) \in T,$$

where k is a positive constant.

6.3.1 Theorem. If, for each $y \in \mathbb{R}^n$, $(y, 0)$ is not a feasible point for (NFEP), then (NFP) and (NFEP) are q-equivalent, where $q(x) = (y, t)$ with $y = tx$ and $t = \phi_0^{-1}(k/g(x))$.

6.3.2 Remarks. Consequently, if either problem has a maximum, then so has the other, and the maxima are equal. The equivalent problem, which generalizes (LFEP), has eliminated the fraction in the objective function. However, conditions must be sought when (NFEP) is convex, or can be replaced by a convex problem, so that effective methods of solution become available.

Proof. If $x \in \mathbb{R}^n$ is feasible for (NFP), then $q(x)$ is defined uniquely, with $t > 0$; since the case $t = 0$ is excluded for (NFEP), the inverse map is given by $x = q^{-1}(y, t) = y/t$, and so q is one-one; $h(x) \in S$ implies $H(y, t) \in T$; and $g(x) > 0$ and $t > 0$ imply $G(y, t) = k$. If (y, t) is feasible for (NFEP), then $q^{-1}(y, t) \in \mathbb{R}^n$ since $t > 0$ by hypothesis; and then $H(y, t) = \psi(t)[h(y/t)] \in T$ implies $h(x) = h(y/t) \in S$. So q maps the constraint set of (NFP) one-one onto the constraint set of (NFEP); and

$$k^{-1}F(y, t) = F(y, t)/G(y, t) = f(x)/g(x).$$

6.3.3 Examples. If $h(x) = Ax - b$ and $\psi(t) = tI$, where I is the identity matrix, then $T = S$ and $H(y, t) = Ay - bt$ is linear. If then $f(x) = c^T x + \alpha$ and $g(x) = d^T x + \beta$, then (LFP) and the linear program (LFEP) are recovered. If instead f is quadratic, thus $f(x) = x^T A x + a^T x + \alpha$ then, with $\phi_0(t) = t^2$, $F(y, t) = y^T A y + t a^T y + \alpha t^2$ is quadratic; if f and g are both quadratic, and h is affine, then (NFEP) has a quadratic objective, and both linear (or affine) and quadratic constraints. A similar result holds with quadratic replaced by a polynomial of degree r, and $\phi_0(t) = t^r$. If f and g are homogeneous functions of degree r, and $\phi_0(t) = t^r$, then $F(y, t) = f(y)$; thus, for example, $f(x) = c^T x + (x^T C x)^{1/2}$, where the matrix C is positive semidefinite, and $r = 1$.

Exercise. Work out in detail the cases where (i) f and g are quadratic, (ii) $f(x) = c^T x + \alpha$ and $g(x) = (x^T B x)^{1/2}$.

6.3.4 Remark. The problem

$$\text{Minimize } f(x) \text{ subject to } x \in E,$$

where the objective function $f(x)$ is nonlinear, can be solved by solving the problem

$$\text{Minimize } t \text{ subject to } t \geqslant f(x) \text{ and } x \in E.$$

Thus the nonlinear objective function is exchanged for a nonlinear constraint.

6.3.5 Lemma. If the constraint set of (NFP) is bounded, then, for each $y \in \mathbb{R}^n$, $(y, 0)$ is not a feasible point for (NFEP).

Proof. By hypothesis, $x^T x \leqslant M < \infty$ for all x feasible for (NFP). For $t > 0$ and $(y, t) \in K$, the constraint set of (NFEP), $x = y/t$ is feasible for (NFP), so $x^T x = t^{-2} y^T y \leqslant M$. If $(y, 0)$ is in the closure of K, then $y^T y \leqslant M t^2 \downarrow 0$ as $t \downarrow 0$, so $y = 0$, hence $0 = G(0, 0) = k > 0$, a contradiction.

6.3.6 Remark. Even if g is a convex function, $G(y, t)$ need not be convex with respect to the vector variable (y, t). For example, consider $g(x) = x^T C x - k$ where C is positive semidefinite and $k > 0$; then g is convex, but taking $\phi_0(t) = t^2$ gives $G(y, t) = y^T C y - k t^2$ which is not convex.

In (NFEP), $t \geqslant 0$ by definition of F and G. Consider the case where (NFP) includes a constraint $x \geqslant 0$. If G is convex, then the set $\{(y, t) : G(y, t) \leqslant k, y \geqslant 0, t \geqslant 0\}$ is convex, but the corresponding set with $G(y, t) = K$ is not generally convex. This suggests modifying the constraint $G(y, t) = k$ of (NFEP) to $G(y, t) \leqslant k$ in order to obtain a convex constraint, but in general this would change the optimum. For the problem

(i) Maximize $(-x - 3)/(x + 1)$ subject to $x \leqslant 2, x \geqslant 0$,

the equivalent problem, but with $= k$ replaced by $\leqslant k$, is

(ii)   Maximize $-y - 3t$ subject to $y + t \leqslant k, y \geqslant 0$,

$$t \geqslant 0, y - 2t \leqslant 0.$$

Here the maximum for (i) occurs at $x = 2$, whereas the maximum for (ii) occurs at $(y, t) = (0, 0)$, which does not correspond to a point of (i). Here also f is a concave function, whereas F is linear (and hence concave). Some conditions when (NFEP) can be modified to have a convex constraint set are given in the two next theorems.

6.3.7 Theorem. Let $p: \mathbf{R}^r \to \mathbf{R}$ be continuous convex; let $W \subset \mathbf{R}^r$. If $p(z)$ attains its maximum on W at $z = z_0$, then the maximum of $p(z)$ for $z \in \overline{\text{co}}\ W$ is also $p(z_0)$.

Proof. Each $z \in \overline{\text{co}}\ W$ is the limit of a sequence of points, each of the form $\Sigma_i(\alpha_i z_i)$, where $z_i \in W$, $\alpha_i \geqslant 0$, $\Sigma_i \alpha_i = 1$. Since p is convex,

$$p(\Sigma_i \alpha_i z_i) \leqslant \Sigma_i \alpha_i p(z_i) \leqslant \Sigma_i \alpha_i p(z_0) = p(z_0).$$

Since p is continuous, $p(z) \leqslant p(z_0)$.

6.3.8 Theorem. In the problem

Maximize $k^{-1} F(y, t)$ subject to $G(y, t) = k, H(y, t) \in T$,

$$y \in \mathbf{R}^n_+, t \in \mathbf{R}_+,$$

denote $z = (y, t)$, and assume that F is convex, $-H$ is T-convex, $G(0) < k$, $Y = \{z \geqslant 0: G(z) = k\}$ is bounded, Y intersects each coordinate axis in $\mathbf{R}^{n+1}_+$, and the sets $E = \{z \geqslant 0: G(z) \leqslant k\}$ and $Q = \{z \geqslant 0: G(z) < k\}$ are convex. Denote $b_j = \inf(Y \cap Z_j)$ where Z_j is the jth positive coordinate axis. Assume that $(y, 0)$ is not feasible, for each $y \in \mathbf{R}^n$. Then the same maximum, at the same point, is obtained when the constraint $G(y, t) = k$ is replaced by

$$G(y, t) \leqslant k \text{ and } \sum_{j=1}^{n} b_j^{-1} y_j + b_{n+1}^{-1} t \geqslant k.$$

6.3.9 Remarks. The problem, thus modified, has a convex constraint set. From the hypotheses, $0 < b_j < \infty$; hence the zero point, which does *not* correspond to a point of (NFP) is excluded. The modified problem is convex if also F is linear.

Proof. Denote $H_+ = \{z \in \mathbb{R}^{n+1} : \Sigma_1^{n+1} b_j^{-1} z_j \geqslant k\}$. The theorem will follow when it is shown that $\overline{co} \, Y = E \cap H_+$. If $z \in H_+$ then, for some $\epsilon > 0$, z is a point of the simplex M whose vertices are the points $(0, 0, \ldots, 0, b_j - \epsilon, 0, \ldots, 0)$, for $j = 1, 2, \ldots, n + 1$. Each vertex in Q; since Q and M are convex, $z \in M \subset Q$, so $z \notin Y$; hence $\overline{co} \, Y \subset E \cap H_+$.

Conversely, let $z \in E \cap H_+$. Since G is continuous, and Y is closed bounded, the ray from 0 through z contains a point $u \in Y$ of maximum norm. If $G(z) = k$ then $z \in Y \subset co(Y)$. If $G(z) < k$, and $u \in [0, z]$, then $G(u) < k$ since E is convex and $G(0) < k$, contradicting $u \in Y$. Hence $z \in L \equiv [0, u]$. Now L intersects the boundary of H_+ in a point w, and z is a convex combination of w, u, so $z \in co \, Y$. Thus $z \in \overline{co} \, Y$.

6.3.10 Example. Consider (NFP) with $S = \mathbb{R}_+^m$, $f(x) = a^T x + \gamma$, $g(x) = d^T x + (x^T C x)^{1/2} + \delta$, and constraints $x \in \mathbb{R}_+^n$ and $h_i(x) = c_i^T x - (x^T D_i x)^{1/2} - \epsilon_i \geqslant 0$ $(i = 1, 2, \ldots, m)$, where C and D_i are positive definite matrices. Assume that this problem has a bounded constraint set. Taking $\phi_0(t) = t$ and $\psi(t) = t$, Theorem 6.3.1 constructs an equivalent problem (NFEP), with a linear objective $F(y, t) = a^T y + \gamma t$; but the constraint

$$G(y, t) \equiv d^T y + (y^Y C y)^{1/2} + \delta t = k$$

does not define a convex set, although G is a convex function. However Theorem 6.3.8 may be applied to construct an equivalent (not q-equivalent) convex problem, noting that F is linear, hence convex, and H is concave.

Exercise. Calculate the added linear constraint for this convex problem.

Exercise. Obtain a dual to the latter convex problem (assuming a constraint qualification holds).

Exercise. Apply Theorem 6.3.8 to the minimization of $f(x) = c^T x/(x^T Cx)^{1/2}$ subject to linear constraints, where C is a positive definite matrix. Show that the constraint set of (NFEP) is here not convex, but there is a convex problem which attains the same maximum at the same point.

Consider the pair of problems

(I):   Maximize $f(x)/g(x)$ subject to $Ax \leqslant b, Qx \leqslant q$;

(II):   Minimize $f(u)/g(u)$ subject to $v \geqslant 0, z \geqslant 0$,

$$b^T v + q^T z \leqslant 0, v^T A + z^T Q = g(u)f'(u) - f(u)g'(u), Qu \leqslant q.$$

Here f and g are real Fréchet differentiable functions of $x \in \mathbb{R}^n$; A and Q are matrices, and b and q are constant vectors. Now assume also that f and g are *homogeneous* functions of the same degree β; *homogeneous* means that, for $t \in \mathbb{R}_+, f(tx) = t^\beta f(x)$, for each $x \in \mathbb{R}^n$. Differentiating with respect to t, then setting $t = 1$, shows that $f'(x)x = \beta f(x)$. Note also that the two constraints $Ax \leqslant b$ and $Qx \leqslant q$ in (I) are treated differently.

6.3.11 Theorem. (homogeneous duality). Let f and g be differentiable, and homogeneous of the same degree β; on the set $H = \{x \in \mathbb{R}^n : Qx \leqslant q\}$, let $-f$ and g be convex functions; let $x \in H$ imply $f(x) \geqslant 0$ and $g(x) \geqslant 0$. Then (II) is a dual to (I).

Proof. Let (I) attain a finite maximum at $x = a$. Since the constraints of (I) are linear, (KT) holds, giving $\lambda \geqslant 0, \mu \geqslant 0$,

$$[g(a)f'(a) - f(a)g'(a)]/[g(a)]^2 = \lambda^T A + \mu^T Q,$$
$$\lambda^T (Aa - b) = 0, \mu^T (Qa - q) = 0.$$

set $\bar{v} = [g(a)]^2\lambda$ and $\bar{z} = [g(a)]^2\mu$. Then $\bar{v} \geqslant 0$, $\bar{z} \geqslant 0$, and

$$g(a)f'(a) - f(a)g'(a) = \bar{v}^T A + \bar{z}^T Q,$$

$$\bar{v}^T b + \bar{z}^T q = [g(a)]^2(\lambda^T A + \mu^T Q)a$$

$$= [g(a)]^2[g(a)f'(a)a - f(a)g'(a)a]$$

$$= [g(a)]^2[g(a)\beta f(a) - f(a)\beta g(a)] = 0.$$

Hence $(u, v, z) = (a, \bar{v}, \bar{z})$ is feasible for (II), and the objective function of (II) at this point equals that of (I) at $x = a$.

Duality will follow if weak duality can be proved. Let x be feasible for (I), and let (u, v, z) be feasible for (II). Then

$$0 \geqslant v^T b + z^T q \geqslant (v^T A + z^T Q)x = g(u)f'(u)x - f(u)g'(u)x.$$

But convexity of g on H gives, for $x, u \in H$, that

$$g(x) - g(u) \geqslant g'(u)(x - u),$$

hence

$$g'(u)x \leqslant g(x) - g(u) + g'(u)u;$$

so $g'(u)x \leqslant g(x) + (\beta - 1)g(u)$. A similar inequality holds for $-f$. Hence, since also $f'(u) \geqslant 0$ and $g(u) \geqslant 0$,

$$0 \geqslant g(u)f'(u)x - f(u)g'(u)x$$

$$\geqslant g(u)[f(x) + (\beta - 1)f(u)] - f(u)[g(x) + (\beta - 1)g(u)]$$

$$= g(u)f(x) - f(u)g(x).$$

Since $x, u \in H$ imply $g(x), g(u) \geqslant 0$, it follows that

$$f(u)/g(u) \geqslant f(x)/g(x),$$

whether finite or infinite.

Remark. The indeterminate case $f(x)/g(x) = 0/0$ is assumed excluded.

Exercise. Show that the function $g(x) = (x^T A x)^{1/2}$, where A is positive semidefinite, is convex, and homogeneous of degree 1. Show that a sum of several such functions is also homogeneous of degree 1.

Exercise. Let $H = \{x \in \mathbb{R}^n : -k \leqslant \xi \leqslant k, \tau = 1\}$, where
$x^T = [\xi^T \quad \tau^T]$ and $\tau \in \mathbb{R}$. Define $f(x) = \tau^T c\tau + 2\tau^T b\xi + \xi^T A\xi$, where the matrix A is negative semidefinite and $c > 0$.
Show that $-f$ is convex on H, and also $f(x) \geqslant 0$ for $x \in H$ if c and k are suitably related.

Exercise. Use Theorem 6.3.11 to find a dual to the problem of minimizing $f(x) = c^T x/(x^T Cx)^{1/2}$ subject to linear constraints, where C is a positive definite matrix, and assuming any necessary hypotheses.

6.4 Algorithms for fractional programming

A linear fractional programming problem reduces to an equivalent linear program, so can be solved by the simplex method. For some nonlinear fractional programming problems there are (see 6.3.8) equivalent *convex* programming problems, which can be solved by standard methods (see Chapter 7). Note, however, that a fractional objective function (to be minimized) is not convex, except in trivial cases; and algorithms for nonconvex problems often find difficulties with convergence, and with more than one local optimum. So it is useful to find an equivalent convex problem, when this can be done. There remain certain algorithms, directly appropriate to fractional programming problems.

A Fréchet differentiable function $\theta : \Gamma \to \mathbb{R}$ is *pseudo-convex* if

$$(\forall x, y \in \Gamma) \quad \theta'(x)(y - x) \geqslant 0 \Rightarrow \theta(y) \geqslant \theta(x).$$

The domain Γ of θ will also be assumed to be convex. Clearly a convex function is pseudoconvex, but not conversely. Now let $\theta(x) = \phi(x)/\psi(x)$ where ϕ and $-\psi$ are convex Fréchet differentiable on Γ, and $\psi(x) > 0$ for each $x \in \Gamma$. Then

$$\theta'(x)(y - x)$$
$$= [\psi(x)\theta'(x)(y - x) - \phi(x)\psi'(x)(y - x)]/[\psi(x)]^2$$
$$\leqslant \{\psi(x)[\phi(y) - \phi(x)] - \phi(x)[\psi(y) - \psi(x)]\}/[\psi(x)]^2$$

$$= [\psi(y)/\psi(x)][\theta(y) - \theta(x)].$$

Hence this function θ is pseudoconvex.

For the problem

$$\text{Minimize } \theta(x) \text{ subject to } x \in \Gamma,$$

where Γ is a bounded polytope (the bounded intersection of finitely many closed halfspaces) in $\mathbb{R}^n$, and $\theta : \Gamma \to \mathbb{R}$ is a continuous pseudoconvex function with continuous first derivative, the algorithm of Frank and Wolfe (1956) is available. (It may also be applied to maximize $\phi(x)/\psi(x)$ subject to $x \in \Gamma$, with $-\phi$ and $-\psi$ convex and differentiable, and ψ positive, since it is then equivalent to minimize the pseudoconvex function $-\phi(x)/\psi(x)$.)

The algorithm constructs a sequence $\{x_r\}$, convergent to the optimum. Start with any $x_0 \in \Gamma$. Since Γ is a polytope, the problem

$$\text{Minimize } \theta'(x_r)v \text{ subject to } v \in \Gamma \qquad (\#)$$

is solved, by linear programming, at $v = v_r$, an extreme point of Γ. Then x_{r+1} is chosen so that

$$\theta(x_{r+1}) = (1 - \rho)\theta/x_r) + \rho \text{ Minimum } \{\theta(x) : x \in [x_r, v_r]\}.$$

Here $0 < \rho < 1$; ρ can vary between iterations but ρ may not tend to zero. A suitable x_{r+1} is obtained by an approximate one-dimensional minimization of $\theta(x)$ along the line segment $[x_r, v_r]$.

Since θ is continuous and Γ compact, the decreasing sequence $\{\theta(x_r)\}$ converges to a finite limit b, and some subsequence $\{z_j\}$ of $\{x_r\}$ converges, say to $\bar{x} \in \Gamma$. Since extr Γ is finite, all v_r for j sufficiently large may be assumed to be the same extreme point, $\bar{v}$ say, of Γ. For $0 < \lambda < 1$,

$$\theta(x_r + \lambda(\bar{v} - x_r)) - \theta(x_r) \geqslant \left[\underset{x \in [x_r, v_r]}{\text{Minimum }} \theta(x) \right] - \theta(x_r)$$
$$\geqslant \rho^{-1}[\theta(x_{r+1}) - \theta(x_r)]$$
$$\to 0 \text{ as } r \to \infty.$$

Hence, considering the subsequence $\{z_j\}$,

$$\theta(\bar{x} + \lambda(\bar{v} - \bar{x})) \geqslant \theta(\bar{x}), \text{ hence } \theta'(\bar{x})(\bar{v} - \bar{x}) \geqslant 0.$$

Now, for any $\xi \in \Gamma$, and all j sufficiently large,

$$\theta'(z_j)(\xi - z_j) \geqslant \theta'(z_j)(\bar{v} - z_j);$$

hence, since θ' is continuous,

$$\theta'(\bar{x})(\xi - \bar{x}) \geqslant \theta'(\bar{x})(\bar{v} - \bar{x}) \geqslant 0.$$

Since θ is pseudoconvex, $\theta(\xi) \geqslant \theta(\bar{x})$.

Note that this problem has a *global* minimum at $\bar{x}$. If the bounded polytope Γ is replaced by a more general convex compact set, then the algorithm still converges, with θ pseudoconvex, assuming that (#) can be solved; $\bar{v}$ is replaced by a convergent sequence in the (compact) closure of extr Γ.

The alternative method of Dinkelbach (1967) makes no convexity hypotheses. Let $\theta(x) = \phi(x)/\psi(x)$, where ϕ and ψ are continuous functions on a compact set S, with $\psi(x) > 0$ for each $x \in S$. Then $\min_{x \in S} \psi(x) > 0$, and $\max_{x \in S} \psi(x) < \infty$. The problem

(I):   Maximize $\phi(x)/\psi(x)$ subject to $x \in S$

is related to the parametric problem

(II):   Maximize $\phi(x) - q\psi(x)$ subject to $x \in S$,

where q is a real parameter; denote the maximum in (II) by $F(q)$. The algorithm depends on the following lemma.

Lemma. F is a strictly decreasing convex continuous function; and Max(I) $= k$ if and only if $F(k) = 0$.

Proof. If $q' < q''$, and $x = \xi$ solves (II) for $q = q''$, then

$$\phi(\xi) - q''\psi(\xi) < \phi(\xi) - q'\psi(\xi) \leqslant F(q'); \text{ hence } F(q'') < F(q').$$

Let x solve (II) for $q = \alpha q' + (1 - \alpha)q''$, where $0 < \alpha < 1$. Then

$$\phi(x) - q\psi(x) = \alpha[\phi(x) - q'\psi(x)] + (1 - \alpha)[\phi(x) - q''\psi(x)]$$
$$\leqslant \alpha F(q') + (1 - \alpha)F(q'').$$

Hence $F(q) \leqslant \alpha F(q') + (1 - \alpha)F(q'')$; so F is a finite convex function on $\mathbb{R}$; hence F is also continuous.

If (I) attains its maximum at $x = \bar{x}$, let $k = \phi(\bar{x})/\psi(\bar{x})$; then $\phi(x) - k\psi(x) \leqslant 0$ for each $x \in S$, and $\phi(\bar{x}) - k\psi(\bar{x}) = 0$, hence $F(k) = 0$. Conversely, if $F(k) = 0$, this argument reverses to show that $\text{Max}(\text{I}) = k$.

The algorithm is as follows. Starting with some $x_0 \in S$ and $q_0 = \phi(x_0)/\psi(x_0)$, a sequence $\{q_r\}$ is obtained iteratively as follows. Suppose that q_r has been calculated. If $F(q_r) = 0$, then the Lemma shows that (I) has been optimized. Otherwise $F(q_r) = \phi(x_r) - q_r\psi(x_r) > 0$, where $x_r \in S$ has been obtained by solving (II) with $q = q_r$; note that the present algorithm assumes that some method is available for optimizing the nonfractional problem (II), for each value of q. Set $q_{r+1} = \phi(x_r)/\psi(x_r)$. Since $\psi(x_r) > 0$ and

$$0 < F(q_r) = \phi(x_r) - q_r\psi(x_r) = (q_{r+1} - q_r)\psi(x_r),$$

the sequence $\{q_r\}$ is strictly increasing, and therefore converges to a limit $q^* \leqslant k$, where $k = \text{Max}(\text{I})$. Since ψ is continuous, and S compact, $\{\psi(x_r)\}$ is a bounded sequence; also F is continuous, and so $\{q_{r+1} - q_r\} \to 0$ implies $F(q^*) = 0$. If $q^* < k < \infty$, then $F(q^*) = 0 = F(k)$, which contradicts the fact that F is strictly decreasing; hence $q^* = k$. If $k = \infty$, a similar argument shows that $\{q_r\} \uparrow \infty$.

Thus $\{q_r\} \uparrow \text{Max}(\text{I})$. Since S is compact, the sequence $\{x_r\}$ has a limit point. Since F is continuous, the maximum of (I) is attained at any limit point of $\{x_r\}$.

6.5 Optimization in complex spaces

The optimization theory for real vector spaces extends as follows to complex spaces. It is presented for spaces of finite dimension, but it extends readily to infinite dimensions. Let V, W, Z be vector spaces, real or complex. Define the inner product $u^*v = u^Tv$ in $\mathbb{R}^k$, or $u^*v = \text{re } u^Hv$ in $\mathbb{C}^k$, where $u^H = \overline{u^T}$ and bar denotes complex conjugate. The inner product is transferred to the vector space V by an isomorphism of V onto $\mathbb{R}^k$ or $\mathbb{C}^k$, for some k. The isomorphism of

$\mathbf{C}^n$ onto $\mathbf{R}^{2n}$, given by $x_j + iy_j \to x_j, y_j \ (j = 1, 2, \ldots, n)$, and the isomorphism of $M = \{(z, z') \in \mathbf{C}^n \times \mathbf{C}^n : z' = \bar{z}\}$ onto $\mathbf{R}^{2n}$, given by $(x + iy, x - iy) \to (2^{1/2}x, 2^{1/2}y)$, both preserve this inner product. The dual cone of a convex cone $S \subset V$ is $S^* = \{y \in V : y^*s \in \mathbf{R}_+ \ (\forall s \in S)\}$; here V is identified with its dual space.

In these terms, Motzkin's alternative theorem (2.5.2) shows that if $A \in L(X, Z), B \in L(X, Y), S \subset Y$ is a convex cone with $\text{int} \, S \neq \emptyset, T \subset Z$ is a closed convex cone, and the cone $A^T(T^*)$ is closed, then exactly one of the systems

$$-A\xi \in T, -B\xi \in \text{int} \, S; p^*B + q^*A = 0, q \in T^*, 0 \neq p \in S^*;$$

has a solution. This is proved for complex vector spaces X, Y, Z by mapping them isomorphically onto real spaces, applying Motzkin's theorem (2.5.2), and then mapping back. This same proof also applies when $X = M$, assuming that A and B are linear with respect to real scalars; although M is not a linear subspace of $\mathbf{C}^{2n}$, the isomorphisms map A and B onto linear mappings between real spaces, to which Motzkin's theorem can be applied.

Let $g : M \to \mathbf{C}^m$ be Fréchet differentiable. Denote

$$g(z, \bar{z}) = g(x + iy, x - iy) = \tilde{g}(x, y).$$

If g were defined and analytic on $\mathbf{C}^n \times \mathbf{C}^n$, then the partial derivatives of g and $\tilde{g}$ would be related by

$$\tilde{g}_x = g_z + g_{\bar{z}} \quad \text{and} \quad \tilde{g}_y = ig_z - ig_{\bar{z}}. \tag{I}$$

The equations (I) will be taken to define g_z and $g_{\bar{z}}$ in case g is only assumed differentiable on M. For each $u \in \mathbf{C}^m, t \in \mathbf{R}$, and derivatives evaluated at $z = a$,

$$g'(a)u = g_z(z - a) + g_{\bar{z}}(\bar{z} - \bar{a});$$

hence, denoting conjugate also by conj,

$$
\begin{aligned}
\overline{g_z}\bar{u} + \overline{g_{\bar{z}}}u &= \text{conj}\,[g_z u + g_{\bar{z}}\bar{u}] \\
&= \text{conj} \lim_{t \to 0} t^{-1}[g(a + tu, \bar{a} + t\bar{u}) - g(a, \bar{a})] \\
&= \lim_{t \to 0} t^{-1}[\bar{g}(a + tu, \bar{a} + t\bar{u}) - \bar{g}(a, \bar{a})] \\
&= \bar{g}_z u + \bar{g}_{\bar{z}}\bar{u}.
\end{aligned}
$$

Here u is arbitrary, hence it follows that

$$\bar{g}_{\bar{z}} = \bar{g}_z \text{ and } \bar{g}_{\bar{z}} = \bar{g}_{\bar{z}}, \tag{II}$$

where $\bar{g}_{\bar{z}}$ means the conjugate of $\partial g/\partial \bar{z}$, whereas $\bar{g}_{\bar{z}}$ means $\partial \bar{g}/\partial \bar{z}$.

This apparatus now allows the Fritz–John theorem, and differentiable duality and converse duality theorems, to be extended from real to complex spaces. Consider the complex problem

(PC): $\underset{\zeta \in M}{\text{Minimize}}$ re $f(\zeta)$ subject to $-g(\zeta) \in S, h(\xi) = 0,$

where $\zeta = (z, \bar{z}) \in M; f : M \to \mathbf{C}, g : M \to \mathbf{C}^m, h : M \to \mathbf{C}^r$ are Fréchet differentiable functions; $S \subset \mathbf{C}^m$ is a convex cone with int $S \neq \emptyset$. For $b = (a, \bar{a}) \in M$, $h'(b)(M)$ is isomorphic (as discussed above) to a real subspace, which is closed since the dimension is finite. This leads to the following complex (FJ) theorem.

6.5.1 Theorem. A necessary condition for (PC) to attain a local minimum at $\zeta = b \in M$ is that there exist $\tau \in \mathbf{R}_+$, $v \in S^*$, $w \in \mathbf{C}^r$, not all zero, such that

(FJC): $\tau(f_z + \bar{f}_{\bar{z}}) + (v^H g_z + v^T \bar{g}_{\bar{z}}) + (w^H h_z + w^T \bar{h}_{\bar{z}}) = 0;$

$$\text{re}(v^H g) = 0;$$

where the functions f, g, h and their partial Fréchet derivatives g_z, etc., are all evaluated at $\zeta = b$.

Proof. If $h'(b)(M) \neq \mathbf{C}^r$, there is nonzero $w \in \mathbf{C}^r$ with $w^* h'(b) = 0$; then (FJC) is satisfied with this w, and $\tau = 0$, $v = 0$. If $h'(b)(M) = \mathbf{C}^r$, then the equivalent in real space of $h(\zeta) = 0$ is locally solvable, hence so is $h(\zeta) = 0$. The Linearization Theorem (2.6.1) applies as well to complex spaces as real spaces. Hence the minimum of (PC) implies that there is no solution $\lambda \in M \times \mathbf{R}$ to the system

$$-A\lambda \in \{0\}, \quad -B\lambda \in \text{int}(\mathbf{R}_+ \times S),$$

where

$$A = [h'(b) \; \vdots \; 0], \quad B = \left[\begin{array}{c|c} f'(b) & 0 \\ \hline g'(b) & g(b) \end{array} \right].$$

Motzkin's alternative theorem then shows that the system $w^*A + u^*B = 0$ has a solution

$$w \in \mathbf{C}^r, 0 \neq u = (\tau, v) \in (\mathbb{R}_+ \times S)^* = \mathbb{R}_+ \times S^*.$$

Hence $v^*g(b) = 0$, and, for each $\zeta = (z, \bar{z}) \in M$,

$$[\tau^*f'(b) + v^*g'(b) + w^*h'(b)]\zeta = 0.$$

Now

$$v^*g'(b)\zeta = \operatorname{re} v^H (g_z z + g_{\bar{z}}\bar{z})$$
$$= \tfrac{1}{2}\{v^H (g_z z + g_{\bar{z}}\bar{z}) + v^T (\overline{g_z z} + \overline{g_{\bar{z}}z})\};$$

and similar expressions hold for $\tau^*f'(b)$ and $w^*h'(b)$. Since $z \in \mathbf{C}^n$ is here arbitrary, the coefficients of z and of $\bar{z}$ must each vanish. This proves the first equation of (FJC); and $v^*g(b) = 0$ proves the second.

Exercise. Obtain alternative, equivalent, expressions by substituting the identities (II) into (FJC).

6.5.2 Corollary. If the constraint system $-g(\xi) \in S$, $h(\zeta) = 0$ is locally solvable, then (KTC) holds, namely (FJC) with $\tau = 1$.

Proof. As in 4.4.3.

6.5.3 Theorem. In (PC), let $\operatorname{re} f$ be $\mathbb{R}_+$-convex and let g be S-convex on M; omit the constraint $h(\zeta) = 0$; let (PC) attain a minimum at $\zeta = b \in M$, and let (KTC) hold at this point (omitting h). Then a dual problem is:

(DC): $\underset{\zeta \in M, v}{\text{Maximize}} \; \operatorname{re} f(\zeta) + v^H g(\zeta)$ subject to $v \in S^*$ and

$$f_z + \overline{f_{\bar{z}}} + v^H g_z + v^T \overline{g_{\bar{z}}} = 0.$$

Proof. As in 4.7, assuming (KTC).

Exercise. Fill in the details of the last two proofs. How is 6.5.3 modified if a linear constraint is included in the primal?

Exercise. Show that the dual constraint can be equivalently written as

$$\overline{f}_z + f_{\bar{z}} + v^H g_z + v^T \overline{g_z} = 0.$$

6.5.4 Theorem. In (PC), let re f be $\mathbb{R}_+$-convex on M, let g be S-convex on M; omit the constraint $h(\zeta) = 0$; let f and g be twice Fréchet differentiable; let $\langle f + \overline{f} + v^H g + v^T \overline{g} \rangle$ be non-singular, where $\langle \phi \rangle$ denotes the matrix

$$\begin{bmatrix} \phi_{zz} & \phi_{z\bar{z}} \\ \phi_{\bar{z}z} & \phi_{\bar{z}\bar{z}} \end{bmatrix}.$$

Then (PC) is a dual problem to (DC).

Proof. The isomorphisms between complex and real spaces map (PC) and (DC) onto equivalent problems in real spaces, to which the real converse duality theorem 4.8.1 can be applied. The constraint $-g(\zeta) \in S$ of (PC) maps to $-g^r(\zeta) \in S^r$, $-g^i(\zeta) \in S^i$, where $g^r = \text{re}\, g$, $g^i = \text{im}\, g$, and S^r and S^i are convex cones in $\mathbb{R}^m$. This pair of inequalities combines to the inequality $-g^\wedge (x, y) \in S^\wedge$ where $S^\wedge = S^r \times S^i$, $z = x + iy$, and $g = ((g^r)^\sim, (g^i)^\sim)$ in the notation of (I). Similarly f maps to a real function $f^\wedge$, and the complex vector v in (DC) maps to a real vector $v^\wedge$. Denote $t = (x, y)$.

The real converse duality theorem (4.8.1) applies if the matrix $D = f^{\wedge \prime\prime}(t) + v^{\wedge T} g^{\wedge \prime\prime}(t)$ is nonsingular. Substitution of the relations (I) between real and complex derivatives shows that, if ϕ is any of f^r, f^i, g^r, g^i, then

$$\begin{bmatrix} \tilde{\phi}_{xx} \\ \tilde{\phi}_{xy} \\ \tilde{\phi}_{yx} \\ \tilde{\phi}_{yy} \end{bmatrix} = \begin{bmatrix} 1 & 1 & 1 & 1 \\ i & -i & i & -i \\ i & i & -i & -i \\ -1 & 1 & 1 & -1 \end{bmatrix} \begin{bmatrix} \phi_{zz} \\ \phi_{z\bar{z}} \\ \phi_{\bar{z}z} \\ \phi_{\bar{z}\bar{z}} \end{bmatrix}. \qquad \text{(III)}$$

Evaluation of the determinant shows that

$$\det \tilde{\phi}''(t) = \det \begin{bmatrix} \tilde{\phi}_{xx} & \vdots & \tilde{\phi}_{xy} \\ \hline \tilde{\phi}_{yx} & \vdots & \tilde{\phi}_{yy} \end{bmatrix}$$

$$= (-4)^n \det \langle \phi \rangle.$$

Now $v^{\wedge T} g^{\wedge ''}(t)$ maps, using (III), onto $\frac{1}{2} v^H \langle g \rangle + \frac{1}{2} v^T \langle \bar{g} \rangle$ in complex space. Therefore D is nonsingular if and only if the complex matrix $\langle f + \bar{f} + v^H g + v^T \bar{g} \rangle$ is nonsingular. Hence (PC) is a dual to (DC) under this hypothesis.

Remark. Complex programming could be applied to electrical networks with alternating currents, with $z \in \mathbf{C}^n$ representing the currents, or voltages, for the n elements of the network.

6.6 Symmetric duality

Let $T \subset \mathbb{R}^n$ and $S \subset \mathbb{R}^m$ be closed convex cones; let $f : \mathbb{R}^n \times \mathbb{R}^m \to \mathbb{R}$ be a twice differentiable function, such that $f(\cdot, y)$ is convex on T for each $y \in S$, and $-f(x, \cdot)$ is convex on S for each $x \in T$. Writing f for $f(x, y)$, and taking vectors in S^* and T^* as row vectors, consider the problems

(PS): Minimize $f - f_y y$ subject to $x \in T$ and $-f_y^T \in S^*$;

(DS): Maximize $f - f_x x$ subject to $y \in S$ and $f_x^T \in T^*$.

These two problems have similar form. The next theorem shows that, under weak conditions, each is dual to the other.

6.6.1 Theorem (symmetric duality). Let (PS) attain a minimum at (x_0, y_0); let int $S^* \neq \emptyset$; let $f_{yy}(x_0, y_0)$ be nonsingular; let the cone

$$[f_y^T \mid f_{yx}^T \mid f_{yy}^T]^T (S),$$

at (x_0, y_0), be closed. Then (DS) is a dual of (PS).

Exercise. From this theorem, formulate similar conditions for (PS) to be a dual of (DS).

Remarks. If the cone S is polyhedral, then the closed-cone condition is automatically satisfied, and the hypothesis int $S^* \neq \emptyset$ may be omitted.

Proof. Applying Theorem 4.4.1 to (PS) at (x_0, y_0) gives $\tau \in \mathbb{R}_+$, $u \in S^{**}$, $v^T \in T^*$, with τ and u not both zero (noting that $x \in T$ is a linear constraint), for which (FJ) holds at (x_0, y_0), namely

(i): $$\tau(f_x - (f_{yx} y)^T) + u^T(f_{yx}^T) - v^T = 0;$$

(ii): $$\tau(f_y - (f_{yy} y)^T - f_y) + u^T f_{yy} = 0;$$

(iii): $$u^T f_y = 0; \quad v^T x = 0.$$

From (ii), $(\tau y - u)^T f_{yy} = 0$; since f_{yy} is assumed non-singular, $\tau y = u$. If $\tau = 0$, then $u = 0$ also; hence $\tau > 0$. So $y = \tau^{-1} u \in S^{**} = S$. Substituting in (i), $f_x = \tau^{-1} v^T \in T^*$. Hence (x_0, y_0) satisfies the constraints of (DS). From (iii), at (x_0, y_0),

$$0 = u^T f_y = \tau y^T f_y \text{ and } 0 = v^T x = \tau f_x x.$$

So the objective functions of (PS) and (DS) are equal at (x_0, y_0).

Also weak duality holds, since

$$[f - f_y y] - [f - f_x x] = f_x x - f_y y \geqslant 0$$

when $x \in T$, $f_x^T \in T^*$, $y \in S$, $f_y^T \in -S^*$. Therefore (DS) is a dual of (PS).

Corollary. Let S and T be polyhedral cones, let one of (PS) and (DS) attain its optimum at (x_0, y_0); let $f_{xx}(x_0, y_0)$ and $f_{yy}(x_0, y_0)$ be nonsingular. Then each of (PS) and (DS) is a dual to the other.

6.6.2 Example. If $f(x, y) = x^T M y + a^T x + b^T y$, where a and b are constant vectors and M is a matrix, then (PS) and (DS) become the pair of linear programs:

Minimize $a^T x$ subject to $x \in T$, $-x^T M \in S^*$;

Maximize $b^T y$ subject to $y \in S$, $y^T M^T \in T^*$.

Exercise. Consider quadratic programs from this standpoint.

Exercise. Prove an analog of theorem 6.6.1 for complex spaces, with $\mathbb{C}$, $\mathbb{C}^m$, $\mathbb{C}^n$ replacing $\mathbb{R}$, $\mathbb{R}^m$, $\mathbb{R}^n$, convex hypotheses on re $f(z, \bar{z}; w, \bar{w})$ (replacing $f(x, y)$), and (PS) modified to:

Minimize $\mathrm{re}\,(f - f_w w - f_{\bar{w}} \bar{w})$ subject to $z \in T$ and
$$- (\mathrm{re}\, f)_{\bar{w}}^T \in S^*.$$

Consider the hypothesis that the matrix

$$\begin{bmatrix} \phi_{zz} & \phi_{z\bar{z}} \\ \phi_{\bar{z}z} & \phi_{\bar{z}\bar{z}} \end{bmatrix}$$

is nonsingular, where $\phi = \mathrm{re}\, f$.

Exercise. Extens the symmetric duality theorem to Banach spaces.

References

Bereanu, B. (1963), Distribution problems and minimum risk solutions in stochastic programming, in *Colloquium on Applications of Mathematics to Economics*, Prékopa, A. (Ed.), Akadémiai Kiadó, Budapest, 37–42.

Bitran, G.R., and Novaes, A.G. (1973), Linear programming with a fractional objective function, *Operations Research*, 21, 22–29.

Charnes, A., and Cooper, W.W. (1962), Programming with linear fractional functionals, *Naval Res. Log. Quart.*, 9, 181–186.

Craven, B.D., and Mond, B. (1972), Converse and symmetric duality in complex nonlinear programming, *J. Math. Anal. Appl.*, 37, 617–626.

Craven, B.D., and Mond, B. (1973), Real and complex Fritz John theorems, *J. Math. Anal. Appl.*, 44, 773–778.

Craven, B.D., and Mond, B. (1973), A Fritz John theorem in complex space, *Bull. Austral. Math. Soc.*, 8, 215–220.

Craven, B.D., and Mond, B. (1973, 1976), The dual of a fractional linear program, *J. Math. Anal. Appl.*, 42, 507–512, and 55, 807.

Craven, B.D., and Mond, B. (1975), On fractional programming and equivalence, *Naval Res. Log. Quart.*, 22, 405–410.

Craven, B.D., and Mond, B. (1976), Duality for homogeneous fractional programming, *Cahiers du Centre d' Études de Recherche Opérationelle*, **18**, 413–417.

Derman, C. (1962), On sequential decisions and Markov chains, *Management Sci.*, **9**, 16–24.

Dinkelbach, W. (1967), On nonlinear fractional programming, *Management Sci.*, **13**, 492–498.

Dorn, W.S. (1962), Linear fractional programming, IBM Research Report RC-830.

Fox, B. (1966), Markov renewal programming by linear fractional programming, *SIAM J. Appl. Math.*, **14**, 1418–1432.

Frank, M., and Wolfe, P. (1956), An algorithm for quadratic programming, *Naval Res. Log. Quart.*, **3**, 95–110.

Gilmore, P.C., and Gomory, R.E. (1961, 1963), A linear programming approach to the cutting stock problem, *Operations Research*, **9**, 849–859; **11**, 863–888.

Isbell, J.R., and Marlow, W.H. (1956), Attrition games, *Naval Res. Log. Quart.*, **3**, 71–94.

Klein, M. (1963), Inspection–maintenance–replacement schedule under Markovian deterioration, *Management Sci.*, **9**, 25–32.

Lintner, J. (1965), The valuation of risk assets and the selection of risky investment in stock portfolios and capital budgets, *Rev. of Econ. and Stat.*, **47**, 13–37.

Mangasarian, O.L. (1969), Nonlinear fractional programming, *J. Opns. Res. Soc. Japan*, **12**, 1–10.

Mangasarian, O.L. (1969), *Nonlinear programming*, McGraw-Hill, New York.

Meister, B., and Oettli, W. (1967), On the capacity of a discrete, constant channel, *Inf. and Control*, **11**, 341–351.

Mond, B., and Craven, B.D. (1973), A note on mathematical programming with fractional objective functions, *Naval Res. Log. Quart.*, **20**, 577–581.

Mond, B., and Craven, B.D. (1975), A class of nondifferentiable complex programming problems, *Math. Operationaforschung und Statistik*, **6**, 577–581.

Mond, B., and Craven, B.D. (1975), Nonlinear fractional programming, *Bull. Austral. Math. Soc.*, **12**, 391–397.

Pollack, E.G., Novaes, G.N., and Frankel, G. (1965), On the optimization and integration of shipping ventures, *International Shipbuilding Progress*, **12**, 267–281.

Schaible, S. (1976), Fractional programming. I, Duality. II, On Dinkelbach's algorithm. *Management Science*, **22**, 858–867, 868–873.

Schaible, S. (1976), Duality in fractional programming: a unified approach, *Operations Research*, **24**, 452–461.

Sharma, I.C. (1967), Feasible direction approach to fractional programming problems, *Opsearch*, **4**, 61–72.

Sharma, I.C., and Swarup, K. (1972), On duality in linear fractional functionals programming, *Zeitschrift für Operations Research*, **16**, 91–100.

Swarup, K. (1965), Programming with quadratic fractional functionals, *Opsearch*, **2**, 23–30.

Ziemba, W.T., Parkan, C., and Brooks-Hill, R. (1974), Calculation of investment portfolios with risk free borrowing and lending, *Management Sci.*, **21**, 209–222.

Notes

Charnes and Cooper (1962) first reduced a linear fractional program to linear programming. The two approaches in 6.2 to duality of a linear fractional program are due respectively to Sharma and Swarup (1972) and to Craven and Mond (1973). For the theory of nonlinear fractional programming, see Craven and Mond (1975, 1977), Mond and Craven (1973, 1975b), Schaible (1976): these papers cite many others on the theory.

The pseudoconvex algorithm of 6.4 was proposed by Frank and Wolfe (1956) for quadratic programming, then shown valid for pseudoconvex functions by Mangasarian (1969). Dinkelbach's parametric method is from Dinkelbach (1967); see also Schaible (1976).

For programming in complex spaces (6.5), refer to Craven and Mond (1972, 1973), Mond and Craven (1975b). The latter paper, which relaxes the differentiability requirement, lists many references by other authors. For symmetric duality, see Craven and Mond (1972), and the references cited there.

CHAPTER SEVEN

Some algorithms for nonlinear optimization

7.1 Introduction

Many algorithms have been proposed for computing the optimum (or optima) of a mathematical programming problem, but there is no universal method. The simplex method for linear programming (3.3) is a highly efficient algorithm; while the number of iterations required to reach an optimum varies widely from one problem to another, the *average* number of iterations, for problems with constraints $Ax - b \in \mathbb{R}_+^m$ with A an $m \times n$ matrix with n much greater than m, is of the order of $2m$, much less than might be expected from the number of vertices of the constraint set. (This remark does not apply to programming restricted to integer values.) If a nonlinear programming problem can be arranged so as to be solvable by a modified simplex method, this is commonly the most efficient procedure. In particular, a problem which allows an adequate approximation by piecewise linear functions, of not too many variables, may be computed as a *separable programming* problem (3.4). Also a problem with a quadratic objective and linear constraints can be solved by a modified simplex method (4.6 and 7.6).

Unless the problem is convex, any iterative method will, if it converges at all, converge to a *local* critical point, which is not necessarily the global optimum (see Fig. 7.1.) This

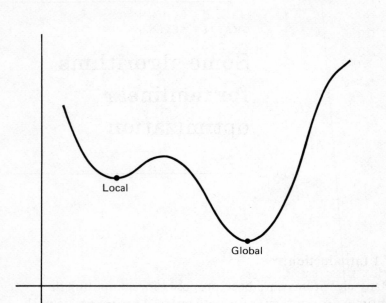

Fig. 7.1. Local and global minima

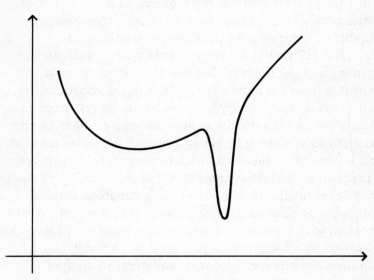

Fig. 7.2. Function with a narrow valley

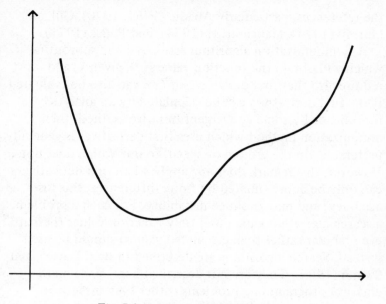

Fig. 7.3. A pseudoconvex function

difficulty does not occur with a convex problem; note that a pseudoconvex objective function (see 6.4), with a convex constraint set, may suffice (e.g. Fig. 7.3). But if a function such as Fig. 7.2 must be minimized, it will be hard to find a computationally efficient algorithm which will find always the global minimum. There is always a tradeoff between efficiency, and the ability to follow small-scale features of the function; in practice, some assumption of smoothness of the function(s) is tacitly made.

In this chapter, the principles of a number of useful minimization algorithms are outlined. The use of these methods will always require a computer, and numerous computer programs exist, for both constrained and unconstrained minimization algorithms. In order to choose between these methods, and find one which is suitable for the class of problems which are to be solved, the user must know at least the principles of a number of the methods. For the fine details needed to write a computer program, the reader is referred to the references, particularly Abadie (1967, 1970), Gill and Murray (1974), Mangasarian (1974), and Polak (1971).

Any minimization algorithm assumes some subroutine which will obtain the function value $f(x)$, given x; and most require also that first derivatives, $f'(x)$, can also be evaluated. If the first derivatives can be calculated by an analytic formula, or by some convergent iterative method, then a minimization method which uses first derivatives is generally preferable, on the ground of speed, to one which does not. However, this remark does *not* apply when first derivatives can only be approximated by finite differences; this loses accuracy, and may produce instability. For such a problem, a search algorithm which uses only function values (perhaps even tabular rather than calculated values) should be used instead. Search algorithms are described in Box, Davies, and Swann (1969). They are not described here, since math-ematical programming problems (other than integer prob-lems) usually assume at least continuous first derivatives, and often second derivatives as well. (Piecewise continuous func-tions with 'corners' occur in minimization problems in some applications; it would be useful to have efficient algorithms for such functions.)

Suppose that an algorithm generates a sequence $\{x_k\}$ of vectors which converges to the desired minimizing vector, $\bar{x}$ say. But convergence is not enough – it may be too slow to be useful. The time for an algorithm to reach an optimum depends both on the rate of convergence, and on the amount of calculation for each *iteration* (the step from x_k to x_{k+1}); these have to be balanced, since improving one will often make the other worse. If, for each k,

$$\|x_{k+1} - \bar{x}\| \leqslant \rho \|x_k - \bar{x}\|, \text{ where } 0 < \rho < 1,$$

then the convergence of $\{x_k\}$ to $\bar{x}$ is called *linear*, or *first order*. If

$$\|x_{k+1} - \bar{x}\| \leqslant \rho_k \|x_k - \bar{x}\|, \text{ where } \{\rho_k\} \to 0,$$

the convergence is *superlinear*. If

$$\|x_{k+1} - \bar{x}\| \leqslant \rho \|x_k - \bar{x}\|^2,$$

the convergence is called *second order*.

The following sections discuss algorithms for minimizing a differentiable real function $f(x)$, where $x \in \mathbb{R}^n$. Unconstrained minimization methods are first discussed (7.2); these are important since a constrained problem can often be solved (7.3) by solving a sequence of unconstrained problems. Some direct methods for unconstrained minimization are given in 7.4; application of the Lagrangean theory of Chapter 4 to constrained minimization algorithms is discussed in 7.5. The remaining sections deal with special methods – quadratic programming by Beale's method (7.6), and the decomposition of a large linear problem into smaller subproblems (7.7).

An optimal control problem may be solvable by representing it as a discrete-time problem (see 4.6), to which finite dimensional methods can be applied. Alternatively, the relevant differential equations can be solved, for a continuous-time problem, with assumed values of certain Lagrange multipliers (compare the discussion in 4.6); these multipliers have then to be adjusted, by some iterative process. Miele (1975) describes such an algorithm. Also the Pontryagin approach can be applied numerically (see 7.5).

As a preliminary, consider *Newton's method* for finding a zero $\bar{x}$ of a real function f of a real variable. The approximation

$$f(x + h) \approx f(x) + f'(x)h = 0 \quad \text{if} \quad h = -[f'(x)]^{-1}f(x)$$

suggests the iterative method

$$x_{r+1} = x_r - [f'(x_r)]^{-1}f(x_r).$$

Suppose, by shift of origin, that $\bar{x} = 0$; suppose that f is twice differentiable; then $f(x_r) \approx f'(0)x_r + \frac{1}{2}f''(0)x_r^2$, and $f'(x_r) \approx f'(0) + f''(0)x_r$; substituting these gives

$$x_{r+1} - \bar{x} \approx q(x_r - \bar{x})^2, \quad \text{where} \quad q = \frac{1}{2}[f'(\bar{x})]^{-1}f''(\bar{x}).$$

Thus Newton's method converges rapidly (second order) if it converges at all. Convergence is guaranteed if the initial

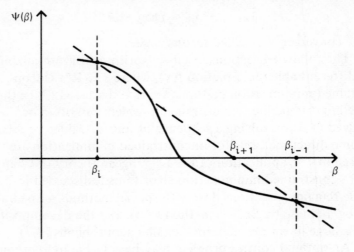

Fig. 7.4. Method of false position

approximation x_0 satisfies $|x_0 - \bar{x}| < |q|^{-1}$; but if x_0 is not so near to $\bar{x}$, some other method must first be used to obtain a suitable starting approximation x_0.

Such considerations often apply to minimization methods. Note also that a minimization method will often require, in each iteration, to minimize $f(x)$ when x moves only along a given line, $x = a + \beta d$ ($\beta \geqslant 0$) say. If the zero $\beta = \bar{\beta}$ is close enough to 0, $\bar{\beta}$ may be found by applying Newton's method to the gradient $\psi(\beta) = f'(a + \beta d)d$, which must vanish at $\bar{\beta}$. But this requires second derivatives of f, and these may not be readily available. Another method is the *false position* method (see Fig. 7.4):

$$\beta_{i+1} = \beta_i - \psi(\beta_i)(\beta_i - \beta_{i-1})/[\psi(\beta_i) - \psi(\beta_{i-1})].$$

It can be shown that, for large i,

$$|\beta_i - \bar{\beta}| \leqslant ab^{c^i}$$

where a, b, c are constants, and $c \approx 1.62$; this assumes that $|\psi''(\beta)/\psi'(\beta)|$ is sufficiently small over an interval of β containing $\bar{\beta}$. Still another method is to approximate the function

$\psi(\beta)$ by a cubic polynomial, and to find the relevant root of this cubic.

7.2 Unconstrained minimization

A differentiable real function f has an unconstrained minimum only where the gradient $\phi(x) = f'(x)$ is zero. A zero $\bar{x}$ of the gradient may be sought by Newton's method; since $\phi(x + h)^T \approx \phi(x)^T + \phi'(x)^T h$, Newton's method takes the form

$$x_{k+1} = x_k - [A(x_k)]^{-1}\phi(x_k)^T,$$

where $A(x_k) = \phi'(x_k)^T = f''(x_k)$ is the *Hessian* of f at x_k; it is assumed that f is twice differentiable, and that the Hessian is invertible. The method will then converge, provided that the initial approximation x_0 is close enough to $\bar{x}$.

A minimum of f could also be sought by the method of *steepest descent*. Starting from an initial approximation x_0 to $\bar{x}$, the method constructs the sequence $\{x_k\}$ by $x_{k+1} = x_k - \alpha_k f'(x_k)^T$, where $\alpha_k > 0$ is chosen so that $f(x)$ decreases along the line $x = x_k - \alpha f'(x_k)^T$ for $0 < \alpha < \alpha_k$, and thereafter increases. A more general descent method is defined by replacing the gradient $f'(x_k)$ by any other vector $-t_k^T$ such that $f'(x_k)t_k < 0$; then $(\partial/\partial\alpha)f(x_k + \alpha t_k) < 0$ at $\alpha = 0$. The *step length* is then given by $f'(x_{k+1})t_k = 0$.

In geometric terms, steepest descent, with $t_k = -f'(x_k)^T$, is suitable when descending the steep sides of the 'valley' formed by the graph of f, where $f(x)$ is well approximated by the tangent plane (or hyperplane). But in the 'bottom' of the valley, near the minimum, a quadratic rather than a linear approximation to f is needed, and steepest descent is then likely to converge slowly, often zigzagging about the minimum instead of moving directly to it. Steepest descent can be shown always to converge (linearly) to the minimum of a quadratic function (assuming $A = f''(x)$ is positive definite); for more general functions f, convergence can also be shown, assuming that $(\forall v \neq 0)\ 0 < mv^T v \leq v^T f''(x)v \leq Mv^T v$ for constants m, M, and x in a suitable region. However, Newton's method is often preferable in the valley region, because of its faster (second order) convergence there, although it only converges when close enough to the minimum.

Various algorithms, therefore, combine features of both Newton and descent methods. Consider therefore a sequence $\{x_k\}$ defined by

$$x_{k+1} = x_k - \alpha_k H_k f'(x_k)^T,$$

where $H_k = I$, or H_k is an approximation to the inverse of the Hessian, or something in between, and α_k is a *step length* determined by some one-dimensional minimization procedure applied to $f(x_k - \alpha H_k f'(x_k)^T)$ for $\alpha > 0$; note also that $f'(x_{k+1})H_k f'(x_k)^T = 0$.

In particular, if $f(x) = \frac{1}{2}x^T A x + b^T x$ with the matrix A positive definite, so that f has a unique minimum, then $g_k \equiv f'(x_k)^T = Ax_k + b$, and $(\partial/\partial\alpha)f(x_k - \alpha H_k f'(x_k)^T) = 0$ when $\alpha_k = g_k^T H_k g_k / (g_k^T H_k^T A H_k g_k)$. Then

$$\frac{(x_{k+1} - \bar{x})^T A(x_{k+1} - \bar{x})}{(x_k - \bar{x})^T A(x_k - \bar{x})} = 1 - \frac{(g_k^T H_k g_k)^2}{(g_k^T H_k^T A H_k g_k)(g_k^T A^{-1} g_k)}$$

$$= 1 - \frac{(w_k^T w_k)^2}{(w_k^T B w_k)(w_k^T B^{-1} w_k)}$$

where S is a symmetric square root of H_k (which is assumed positive definite symmetric), $B = S^T A S$, $w_k = Sg_k$,

$$= 1 - \{\Sigma_1^n \lambda_i u_i^2 \cdot \Sigma_1^n \lambda_j^{-1} u_j^2\}^{-1}$$

where $\lambda_1 \geqslant \lambda_2 \geqslant \ldots \geqslant \lambda_n$ are eigenvalues of B (and so of $H_k A$) and $\Sigma_1^n u_i^2 = 1$, by expanding B in terms of eigenvalues and eigenvectors,

$$\leqslant 1 - \frac{4\lambda_1 \lambda_n}{(\lambda_1 + \lambda_n)^2} = \left\{\frac{\lambda_1 - \lambda_n}{\lambda_1 + \lambda_n}\right\}^2$$

on maximizing with respect to the u_i. This *convergence rate inequality* (CRI) indicates the convergence rate expected for more general functions which may be approximated by a quadratic near the minimum $\bar{x}$, provided that H_k is a reasonable approximation to the matrix $[f''(x_k)]^{-1}$ when $\{x_k\} \to \bar{x}$.

But if $H_k A$ is *ill-conditioned*, thus if λ_1/λ_n is large, then the convergence will be slow.

In order to discuss suitable descent directions, define two directions t_1 and t_2 as *conjugate* with respect to the real symmetric matrix A if $t_1^T A t_2 = 0$. In particular, an orthonormal system of eigenvectors for A has this property. For the above iterative method, g_k and g_{k+1} are conjugate with respect to H_k. For $f(x) = \frac{1}{2}x^T A x + b^T x$, let $t_1, t_2, \ldots, t_n$ be n mutually conjugate directions with respect to A. Define a descent sequence by $x_{k+1} = x_k + \alpha_k t_k$, with α_k such that $f'(x_{k+1})t_k = 0$. Then, for $j + 1 < i$,

$$g_i^T t_j = (Ax_i + b)^T t_j = \left(b + Ax_{j+1} + \sum_{k=j+1}^{i-1} \alpha_k A t_k\right)^T t_j$$

$$= g_{j+1}^T t_j + \sum_{k=j+1}^{i-1} \alpha_k t_k^T A t_j = 0 + 0.$$

Hence $g_i^T t_j = 0$ for $j < i$. Consequently the minimum of a positive definite quadratic function can be found by n descent steps along conjugate directions, since g_{n+1} is orthogonal to the n linearly independent vectors $t_1, \ldots, t_n$, and thus $g_{n+1} = 0$; since f is convex, this implies that x_{n+1} is a minimum.

A *conjugate gradient* algorithm uses a descent sequence with directions t_k generated by $t_{k+1} = \beta_{k+1} t_k - g_{k+1}$, where $g_{k+1} \equiv f'(x_{k+1})^T$ and β_{k+1} is chosen as $g_{k+1}^T A t_k/(t_k^T A t_k)$, so that $t_{k+1}^T A t_k = 0$. Consider a quadratic $f(x) = \frac{1}{2}x^T A x + b^T x$; if $t_r^T A t_s = 0$ for $1 \leqslant r < s \leqslant k$, then

$$t_r^T A t_{k+1} = \beta_{k+1} t_r^T A t_k - (A t_r)^T g_{k+1} = 0 - 0$$

since $A t_r$ is in the subspace spanned by $t_1, t_2, \ldots, t_k$ and $t_j^T g_{k+1} = 0$ for $j \leqslant k$, by the previous paragraph. Hence the directions $\{t_k\}$ are mutually conjugate directions with respect to A (Luenberger, 1969).

These results for quadratic functions may be applied, modified a little, to nonquadratic functions f, noting that a twice-differentiable function is well approximated by a quadratic, near its minimum $\bar{x}$; $f''(\bar{x})$ is assumed positive definite. An iteration is described by

$$x_{k+1} = x_k + \alpha_k t_k$$

where α_k minimizes $f(x_k + \alpha t_k)$ with respect to α,

then $t_{k+1} = \beta_{k+1} t_k - g_{g+1}$ (with $g_{k+1} = f'(x_{k+1})^T$).

While the formulas given above for $A = f''(x_k)$, this approach requires the considerable extra computation of the Hessian matrix $f''(x_k)$. However, the equation $g_i^T t_j = 0$ for $j < i$, together with the fact that g_j is in the subspace spanned by $t_1, t_2, \ldots, t_j$, may be used to obtain another expression for β_{k+1}, namely $\beta_{k+1} = \|g_{k+1}\|^2 / \|g_k\|^2$, which does not involve A.

The Fletcher–Reeves method obtains α_k by a one-dimensional minimization of $f(x_k + \alpha t_k)$ with respect to α, using one of the methods given in 7.1, and then calculates β_{k+1} from the formula from g_k and g_{k+1}, without requiring the Hessian. The method is commenced with any initial vector x_0, and initial direction $t_0 = -f'(x_0)^T$. In order to obtain a good approximation to conjugate directions, it is desirable to reset $t = -f'(x)^T$ after each n iterations. Since this amounts to a steepest descent step, convergence is assured, assuming the positive definite hypothesis on $f''(x)$. If each block of n iterations in the conjugate gradient method with reset is regarded as a 'super-iteration', then the super-iterations converge superlinearly to a local minimum, subject to some restrictions on the directions t_k (McCormick and Pearson, 1969).

These methods differ in their rates of convergence to the minimum $\bar{x}$. Newton's method may be written as $x_{k+1} = F(x_k)$, where $F(x) = x - [f''(x)]^{-1} f'(x)^T$. Then $F(\bar{x}) = \bar{x}$; and if I is the identity matrix and h is the derivative of $[f''(x)]^{-1}$ at $\bar{x}$, then

$$F'(\bar{x}) = I - h f'(\bar{x})^T - [f''(\bar{x})]^{-1} f''(\bar{x}) = I - 0 - I = 0.$$

Hence, assuming that F has a bounded second derivative,

$$\|x_{k+1} - \bar{x}\| = \|F(x_k) - F(\bar{x})\|$$
$$= \|F'(\bar{x})(x_k - \bar{x}) + \tfrac{1}{2}(x_k - \bar{x})^T B(x_k - \bar{x})\|$$

$$= \|\tfrac{1}{2}(x_k - \bar{x})^T B(x_k - \bar{x})\|,$$

where $B = F''(x)$ for some x in the line segment $[\bar{x}, x_k]$. Hence, if $\|x_k - \bar{x}\|$ is small enough, then $\|x_{k+1} - \bar{x}\| \leqslant \tfrac{1}{2}c\|x_k - \bar{x}\|^2$, where c is an upper bound to $\|F''(x)\|$ for x near $\bar{x}$, and therefore depends on the third derivative of f. Hence Newton's method converges rapidly (second order), if it converges at all.

In contrast, the CRI for the modified descent method shows that $\|x_{k+1} - \bar{x}\|_A \leqslant \rho\|x_k - \bar{x}\|_A$, where $0 < \rho < 1$ and $\|y\|_A = (y^T A y)^{1/2}$; then $\|y\|_A$ is a norm if A is positive definite. Thus the modified descent method gives linear convergence. Of course, for a given objective function f, a method may happen to converge faster than these upper bounds indicate. For the Fletcher–Reeves method with resetting, the CRI may be applied to a block of n iterations. It has otherwise been shown that if the set $\{x : f(x) \leqslant f(x_0)\}$ is bounded, then the sequence $\{x_k\}$ has a limit point at which $f'(x)$ is zero. (An alternative formula

$$\beta_{k+1} = (g_{k+1} - g_k)^T g_k / [(g_{k+1} - g_k)^T t_k]$$

is also consistent with conjugate directions, and some computational evidence suggests that it may be preferable to the earlier formula.)

The Fletcher–Reeves conjugate gradient method with resetting requires less information to be stored than those methods which use second derivatives; and it has been described as extremely reliable (Fletcher, in Murray, 1972). But it often requires more iterations to converge than does the DFP method, next to be described (see Polak, 1971).

Another approach is to generate a sequence $\{H_k\}$, suitably approximating the inverse Hessian $[f''(x_k)]^{-1}$; note that here the Hessian is not calculated directly. Here, an iteration consists of a descent step in the direction $t_k = -H_k f'(x_k)^T$, followed by an updating of H_k to H_{k+1}, ready for the next iteration. Desirably, the updating should make the directions t_i conjugate, or approximately so. To obtain the Davidson–Fletcher–Powell (DFP) method, consider the updating formula

$$H_{k+1} = H_k + \gamma_k s_k s_k^T + \delta_k H_k y_k y_k^T H_k,$$

where $s_k = x_{k+1} - x_k$, $g_k = f'(x_k)^T$, $y_k = g_{k+1} - g_k$, and H_k is assumed symmetric. The requirement that the t_i be conjugate with respect to the matrix A, when f is quadratic, is fulfilled if $H_i A t_j = \gamma_j t_j$ whenever $j < i$, or equivalently if $H_i y_j = \epsilon_j s_j$ for $j < i$, where the ϵ_j are constants. Rewrite the updating formula as

$$H_{i+1} = H_i + \bar{\gamma}_i \frac{s_i s_i^T}{s_i^T y_i} - \bar{\delta}_i \frac{H_i y_i y_i^T H_i}{y_i^T H_i y_i}.$$

Now, for $j < i$, $s_i^T y_j = \alpha_i t_i^T (g_{j+1} - g_j) = 0$ if the directions t_j are conjugate for $j \leqslant i$, and similarly

$$y_i^T H_i y_j = \epsilon_i s_i^T y_j = \epsilon_i \alpha_i t_i^T y_j = 0.$$

Hence the updating formula gives $H_{i+1} y_j = \epsilon_j s_j + \bar{\gamma}_i 0 + \bar{\delta}_i 0$. So, by induction, the desired conjugate property holds for all i.

The DFP algorithm (also called *variable metric* algorithm) sets the constants $\bar{\gamma}_i$ and $\bar{\delta}_i$ all equal to 1; and initially $H_0 = I$. If f is quadratic, with positive definite Hessian A, then the DFP method gives the minimum in n iterations, and the t_k are conjugate gradients. But, although $H_k \to A^{-1}$, H_k can be ill-conditioned compared to A, and thus the convergence can be slower than λ_1/λ_n for A would indicate (using the CRI).

Assume now that f is twice continuously differentiable, and that $f''(x)$ is positive definite in a neighbourhood of the minimum $\bar{x}$. If, in each iteration, the one-dimensional minimization is done *exactly*, then it has been shown that the convergence rate of DFP is superlinear, and each H_k is positive definite. Computational experience (see Broyden, 1972) shows, in fact, that the convergence rate of DFP deteriorates appreciably unless the step length α_k at each iteration is quite accurately determined. This fact can be partly explained in terms of eigenvalues (compare the CRI). Given, however, an accurate one-dimensional minimization in each iteration, DFP has been found in practice to be a widely applicable and efficient minimization method.

DFP is called a *rank two* method, since H_k is updated by adding two symmetric matrices of rank one. Various other methods, of rank two or rank one (i.e. adding a matrix of rank 1) have been described (see Murray, 1972).

7.3 Sequential unconstrained minimization

Suppose that the constrained minimization problem

(P):   Minimize $f(x)$ subject to $k(x) \in \mathbb{R}^m_+$,

where $f : \mathbb{R}^n \to \mathbb{R}$ and $k : \mathbb{R}^n \to \mathbb{R}^m$ are continuous, attains a local minimum v at $x = \bar{x}$. Let $k(x)_-$ denote $k(x)$ if $k(x) \notin \mathbb{R}^m_+$, or 0 if $k(x) \in \mathbb{R}^m_+$. For some $e \in \text{int } \mathbb{R}^m_+$ and some $\epsilon > 0$, assume that the set $B = \{x \in \mathbb{R}^n : k(x) + \epsilon \in \mathbb{R}^m_+,$ $f(x) \leqslant v + \epsilon\}$ is compact. For a sequence of positive parameters $\{t_i\} \uparrow \infty$, consider the unconstrained minimization problem

(P$_i$):   Minimize $V(x, t_i) = f(x) + t_i \|k(x)_-\|^2$.

As in Fiacco and McCormick (1968) assume that, for the chosen ϵ and e, and whenever t_i is large enough, (P$_i$) attains an *unconstrained* minimum at a point x_i in the compact set B. Note however that x_i need not be a feasible point for (P); consequently $f(x)$ must be defined for points x outside the constraint set for (P). If (P$_i$) is solved for a sequence of values of t_i, tending to ∞, then the *penalty function* $t_i \|k(x)_-\|^2$ forces x_i to approach the constraint set of (P). This suggests that $\{x_i\}$ will converge to an optimal solution of (P).

Assume that (P) attains a finite minimum at $x = \xi$. Let η be any limit point of $\{x_i\}$; one exists since B is compact; then a subsequence $\{x_{i_j}\} \to \eta$. If $k(\eta) \notin \mathbb{R}^m_+$, then $\|k(\eta)_-\| > 0$, so $\{V(x_{i_j}, t_{i_j})\} \to \infty$ as $j \to \infty$, contradicting $V(x_i, t_i) \leqslant V(\xi, t_i) = f(\xi) < \infty$ for all i; therefore $k(\eta) \in \mathbb{R}^m_+$. If (P) does not attain a local minimum at η, then

$$\lim_{j \to \infty} V(x_{i_j}, t_{i_j}) = f(\eta) > f(\xi') = \lim_{i \to \infty} V(\xi', t_i)$$

for some minimum ξ' of (P); and this contradicts the definition of $V(x_i, t_i)$ as a minimum, for large enough i. It follows that each convergent subsequence of $\{x_i\}$ converges to some local minimum of (P).

Fiacco and McCormick (1968) call this method an *exterior-point* (penalty) algorithm because an optimum is approached by (non-feasible) points, exterior to the constraint set of (P). If the constraints of (P) are instead $g_s(x) \leqslant 0$ $(s = 1, 2, \ldots, p)$ and $h_j(x) = 0$ $(j = 1, 2, \ldots, q)$, then the penalty function becomes

$$t_i \left\{ \sum_{s=1}^{p} [g_s(x)_+]^2 + \sum_{j=1}^{q} [h_j(x)]^2 \right\} = t_i \{ \|g(x)_+\|^2 + \|h(x)\|^2 \}$$

where $g_s(x)_+ = g_s(x)$ if $g_s(x) \geqslant 0$, or 0 otherwise. The preceding theory applies, with this modified penalty function, and B redefined as

$$\{ x \in \mathbb{R}^n : -g(x) + \epsilon \in \mathbb{R}_+^p, h(x) = 0, f(x) \leqslant v + \epsilon \}.$$

A different penalty function $t_k \phi(x)$ could be used, provided that ϕ is continuous, $\phi(x) = 0$ for all feasible x, and $\phi(x) > 0$ for all x not feasible for (P).

This method is often very effective. However, the graph of $V(x, t)$ often forms an increasingly steep-sided valley as t becomes large, so that the minimization of (P_k) may require a great deal of computation; note also that the Hessian may become ill-conditioned. Also there is no obvious way of choosing how fast $\{t_i\} \uparrow \infty$, except by observing when $\|x_{i+1} - x_i\|$ becomes small.

For a constraint system $k_s(x) \geqslant 0$ $(s = 1, 2, \ldots, m)$, there is an alternative approach of minimizing

$$F(x, r) = f(x) + r \sum_{s=1}^{m} \phi_s(k_s(x))$$

for a sequence of values of $r \downarrow 0$, where each $\phi_s(t) \to \infty$ as $t \downarrow 0 +$. This method approaches an optimum through feasible points of (P); the function added to $f(x)$ is called an *interior-point* penalty function or *barrier function*, whose effect is to prevent the point from leaving the constraint set. However, the barrier function method requires the constraint set to have interior points, so that equality constraints are here

excluded. Possible functions $\phi(t)$ are t^{-1} or $-\log t$. Convergence of the method, given f and g_s continuous, may be proved
similarly to the exterior-point algorithm. It is also possible
to use a penalty function which combines both an exterior-
point penalty, and an interior-point barrier function, the
latter for the inequality constraints only.

For the interior-point method, assume that f and k in (P)
are twice continuously differentiable, with f and each $-k_s$
convex. Let $\phi_s(y)$ have derivative $y^{-\nu}$ where ν is a positive
integer. Let $F(x, r)$ reach its minimum at $x = x(r)$, for $r > 0$.
Lootsma (1969) has shown that $u_s(r) = r[k_s(x(r))]^{-\nu}$ tends
to a limit $\bar{u}_s$ as $r \downarrow 0$; and that the Hessian matrix of $F(x, r)$,
evaluated at $(x(r), r)$, takes the form

$$\left[f''(x(r)) - \sum_{s=0}^{m} u_s k_s''(x(r)) \right] + r^{-1/\nu} \left[\nu \sum_{s=1}^{m} u_s^{1+1/\nu} k_s'(x)^T k_s'(x) \right]$$

$$= A + r^{-1/\nu} B, \text{ say.}$$

As $r \downarrow 0$, it follows that this Hessian has q eigenvalues which
tend to ∞ as $r^{-1/\nu}$, the rest remaining finite, where q is the
rank of the matrix B. Since non-binding constraints do not
affect the convergence, they may be omitted here, and q
replaced by the number of binding constraints, assuming that
the latter have linearly independent gradients at the minimum.
Under the same hypotheses, including continuous second
derivatives of f and k_s, it has been shown also that $x(r)$ is a
continuously differentiable function of $r^{1/\nu}$ as $r \downarrow 0$. Since
$x(r_i)$ is only used as a starting value for computing $x(r_{i+1})$, it
may not be necessary to evaluate $x(r_i)$ very accurately.
Alternatively, an extrapolation procedure may be based on
accurate evaluations of $x(r)$ for a few values of r, using
$x(r) \approx \bar{x} + cr^{1/\nu}$ $(r \downarrow 0)$.

There is a choice of algorithms for performing the sequence
of unconstrained minimizations required in these methods –
see 7.2. Often the DFP method is suitable.

Consider now the constraints $g_s(x) \leqslant 0$ $(s = 1, 2, \ldots, p)$
and $h_j(x) = 0$ $(j = 1, 2, \ldots, q)$, and the modified penalty
function

$$P(x, t) = f(x) + t\left\{\sum_{s=1}^{p} g_s(x)_+ + \sum_{j=1}^{q} |h_j(x)|\right\}.$$

Assume that the problem attains a minimum at $x = \bar{x}$, and that (KT) holds there. Pietrzykowski (1969) has shown then that $P(x, t)$ attains a local unconstrained minimum at $x = \bar{x}$ for sufficiently large, but *finite*, t. This *exact minimum penalty algorithm* thus avoids the limit $t \to \infty$, and the consequent ill-conditioning, but at the expense of a function P which is only piecewise differentiable. So special methods are needed for its minimization, and it is not yet clear what method would be efficient here.

7.4 Feasible direction and projection methods

Some of the unconstrained methods of 7.2 have been adapted to unconstrained minimization. Their effectiveness then depends largely on how well they can follow the boundary (in general curved) of the constraint region, without requiring excessively small steps (often with a 'zigzag' path), or converging to some point which is not a local minimum.

For the constrained problem

Minimize $f(x)$ subject to $x \in K$,

a *feasible direction* d_r at the point $x_r \in K$ is a direction such that $x_r + \alpha d_r \in K$ whenever $0 \leqslant \alpha \leqslant \delta$, for some $\delta > 0$. Since usually x_r is a boundary point of K, not all directions from x_r are feasible.

Consider first linear constraints: $x \in K$ iff $Ax \leqslant b$; suppose that the binding constraints are described by a submatrix A_1 of A, thus $A_1 x_r = b$. An appropriate feasible direction may be found (see Zoutendijk, 1960) by solving a subsidiary linear program:

Minimize$_d$ $f'(x_r)d$ subject to $A_1 d \leqslant 0$, $|d| \leqslant e$.

Here $|d|$ is the vector with components $|d_i|$; $e^T = (1, 1, \ldots, 1)$. Then set $x_{r+1} = x_r + \alpha_r d_r$ where $\alpha = \alpha_r$ minimizes $f(x_r + \alpha d_r)$ over $\alpha \in (0, 1]$.

In general, there may be no feasible directions; and the iteration step may have to be modified to make $\{x_r\}$ converge. Suppose now that K is compact convex, specified by constraints $g_i(x) \leqslant 0$, not necessarily linear, and that f is continuously differentiable. The subsidiary linear program may be modified to

$$\underset{d,\mu}{\text{Minimize }} \mu \text{ subject to } f'(x_r)d \leqslant \mu, g(x_r) + g'(x_r)d \leqslant \mu e,$$

$$|d| \leqslant \mu e.$$

The feasible direction method, with this subsidiary linear program, has been shown to converge, at the expense of a great deal of linear programming. (Of course, the optimum for the last linear program may be used as an initial basis for the next.) The linear program constraints may be replaced by

$$g_i'(x_r)d \leqslant \mu \text{ whenever } -\epsilon_r \leqslant g_i(x_r) \leqslant 0,$$

where μ is here replaced by 0 if g_i is affine, and ϵ_r is a suitable tolerance. If $\epsilon_r < \mu_{\text{optimal}}$, set $\epsilon_{r+1} = \frac{1}{2}\epsilon_r$. Assume that the first derivatives of g are Lipschitz continuous, meaning that

$$\|g'(x) - g'(y)\| \leqslant \text{const.} \|x - y\|$$

for all feasible x and y, and assume that f is continuously differentiable. For this version of Zoutendijk's feasible direction method, it can be shown that each limit point of the sequence $\{x_r\}$ satisfies the (FJ) condition for the minimization problem. Under suitable additional convexity conditions (see 4.5), this implies a minimum. This convergence does *not* require that the one-dimensional minimization for α_r is done exactly. However, equality constraints, if present, must be linear.

Various *gradient projection* methods use the feasible direction idea, but try to reduce the calculation at each iteration. If the constraints are *linear*, $Ax \leqslant b$, then a feasible direction d_r is sought, which satisfies $f'(x_r)d_r < 0$ and $A_1 d_r = 0$. While this is more restrictive than $A_1 d_r \leqslant 0$ previously considered, it allows a suitable vector d_r to be obtained as the *projection* of $-f'(x_r)^T$ onto the tangent subspace defined by the binding constraints. Then α_r is obtained as before by minimizing f along the line $x_r + \alpha d_r$. For linear constraints, the projection matrix required to calculate d_r can be updated from one iteration to the next, so does not need a complete new calculation. While there is no general convergence proof, this gradient projection method, for linear constraints, has shown itself effective in practice; note that it does *not* require one linear program per iteration.

For nonlinear constraints, the boundary of the constraint set is usually curved. A direction d_r can still be calculated, by projecting the gradient $-f'(x_r)^T$ onto the tangent space to the surface defined by those constraints, say $\phi(x_r) = 0$, which are binding at x_r. The required projection matrix is $P = I - A(A^T A)^{-1} A^T$, where $A = \phi'(x_r)$. Note that $A^T A$ is assumed to have an inverse; also, since A depends on x_r, P must here be calculated afresh for each iteration. With a curved boundary $d_r = P(-f'(x_r)^T)$ is not necessarily a feasible direction, and $x_r + \alpha_r d_r$, with α_r chosen to minimize f, must be corrected to be a feasible point, by setting $x_{r+1} = x_r + \alpha_r d_r + \phi'(x_r)v$ for a suitable (small) term v. An iterative scheme convergent to x_{r+1} has been given (Luenberger, 1973) as

$$z_{i+1} = z_i - (I - P)\phi(z_i) \ (i = 0, 1, \ldots); \quad z_0 = x_r + \alpha_r d_r.$$

A linear rate of convergence can be proved for the gradient projection method, making sufficient assumptions. While the method is often efficient, the computing details, and variations, required to ensure convergence are complicated – see Chapter 5 of Gill and Murray (1974).

7.5 Lagrangean methods

For the problem with equality constraints

(EP):   Minimize $f(x)$ subject to $h(x) = 0$,
$\quad\quad\quad\quad\quad x$

assume that the minimum is attained at $x = \bar{x}$, and that f and g are twice continuously differentiable. If $h(x) = 0$ is locally solvable at $\bar{x}$, then (KT) holds, with optimal Lagrange multiplier $\bar{\lambda}$ say, and so to each $d \in S \equiv h'(\bar{x})^{-1}(0)$ corresponds a local solution $x = \bar{x} + \alpha d + o(\alpha)$ to $h(x) = 0$. The Lagrangian is $L(x, \lambda) = f(x) - \lambda h(x)$. If also $M \equiv (\partial^2/\partial x^2)L(\bar{x}, \bar{\lambda})$ is positive definite on S, then for small enough α,

$$f(x) = L(x, \bar{\lambda}) = L(\bar{x}, \bar{\lambda}) + (\partial/\partial x)L(\bar{x}, \bar{\lambda})\alpha d$$
$$+ \tfrac{1}{2}\alpha^2 d^T M d + o(\alpha^2) > L(\bar{x}, \bar{\lambda})$$

by the positive definiteness, since $(\partial/\partial x)L(\bar{x}, \bar{\lambda}) = 0$ by (KT). Thus the problem (EP) has an isolated minimum point.

Consider the related unconstrained minimization problem

(UP): $\underset{x}{\text{Minimize }} F(x, \lambda) \equiv f(x) + \tfrac{1}{2}\mu\|h(x)\|^2 - \lambda h(x),$

where μ is a fixed positive parameter, and $\|u\| \equiv (u^T u)^{1/2}$. Note that $F(x, \lambda)$ differs from $f(x) + \tfrac{1}{2}\mu\|h(x) - \mu^{-1}\lambda\|^2 \equiv \tilde{F}(x, \lambda)$ only by the added term $\tfrac{1}{2}\mu^{-1}\|\lambda\|^2$; it turns out that F and $\tilde{F}$ lead to the same iterative methods. Denote $F'(x, \lambda) = (\partial/\partial x)F(x, \lambda)$, and $F''(x, \lambda) = (\partial^2/\partial x^2)F(x, \lambda)$. Using (KT) for (EP),

$$F'(\bar{x}, \bar{\lambda}) = f'(\bar{x}) - \bar{\lambda}h'(\bar{x}) + \mu h(\bar{x})^T h'(\bar{x}) = 0.$$

Note also that $F = L$ plus a penalty term for the constraint $h(x) = 0$.

For (UP), the Hessian $F''(\bar{x}, \bar{\lambda}) = L''(\bar{x}, \bar{\lambda}) + \mu h'(\bar{x})^T h'(\bar{x})$ since $h(\bar{x}) = 0$; suppose that the first term, M, in this Hessian is positive definite on S; the second term is positive definite on the subspace orthogonal to S; so, for large enough μ, $F'(\bar{x}, \bar{\lambda})$ is positive definite. By continuity, so is $F''(x, \lambda)$ for x near $\bar{x}$ and λ near $\bar{\lambda}$. Since also $F'(\bar{x}, \bar{\lambda}) = 0$, it follows that $x = \bar{x}$ minimizes $F(x, \bar{\lambda})$. Also the equation $F'(x, \lambda) = 0$ is locally solvable, for λ near $\bar{\lambda}$, by a unique $x = x(\lambda)$, since $F''(x, \lambda)$ is invertible; denote $Q(\lambda) = F(x(\lambda), \lambda)$. Now $Q(\lambda)$ is to be stationary; this occurs at $\lambda = \bar{\lambda}$. Routine calculations then show that $Q'(\lambda) = -h(x(\lambda))$, and that $-Q''(\lambda)$ is positive definite if $F''(x, \lambda)$ is, with $Q''(\lambda)$ approximately equal to $-\mu I$ for large positive μ, where I is the unit matrix.

It is therefore appropriate to apply a modified Newton method to the calculation of $\bar{\lambda}$, thus by iterating

$$\lambda_{k+1} = \lambda_k - \mu^{-1} h(x_k)^T \quad (k = 1, 2, \dots).$$

If each $x_k = x(\lambda_k)$ is readily calculable, then this iterative method converges rapidly, since the eigenvalues of $Q''(\lambda)$ are close together. In practice, x_k is taken instead as the value of x at which $F(x, \lambda_k)$ attains its unconstrained minimum value; this is a penalty-function minimization problem.

Consider now the inequality-constrained problem

(IP): $\underset{x}{\text{Minimize}} f(x)$ subject to $g(x) \geqslant 0$.

For (IP), $\tilde{F}$ may be replaced, as in Rockafellar (1974), by

$$\tilde{\Phi}(x, \lambda) = f(x) + \tfrac{1}{2}\mu \|(g(x) - \mu^{-1}\lambda)_-\|^2$$
$$= f(x) + \tfrac{1}{2}\mu \|(-g(x) + \mu^{-1}\lambda)_+\|^2,$$

in which $y_+ = (y_{1+}, \dots, y_{m+})$, $y_{i+} = \max\{y_i, 0\}$, $y_- = (-y)_+$, for $y \in \mathbb{R}^m$. This choice of penalty function leads to results similar to those for equality constraints. The iterative method, obtained above for equality constraints, is modified to

$$\lambda_{k+1} = (\lambda_k - \mu^{-1} g(x_k))_+ \quad (k = 1, 2, \dots),$$

in which $x = x_k$ is an unconstrained minimum point of $\tilde{\Phi}(x, \lambda_k)$, or of $\Phi = \tilde{\Phi} - \tfrac{1}{2}\mu^{-1}\|\lambda\|^2$. The good convergence behaviour for the equality case extends to inequalities. Note that the penalty term in Φ is the sum of components $-\lambda_i b_i + \tfrac{1}{2}\mu b_i^2$ if $b_i \leqslant \lambda_i/\mu$, or $-\tfrac{1}{2}\lambda_i^2/\mu$ if $b_i \geqslant \lambda_i/\mu$, where $b_i = g_i(x)$ $(i = 1, 2, \dots, m)$. The cone here is $\mathbb{R}_+^m$.

In the penalty function methods of 7.3, the parameter $t_k \to \infty$, which often leads to ill-conditioned problems. In contrast, the present Lagrangean method requires μ sufficiently large, but μ need *not* tend to ∞; hence the ill-conditioning may be avoided, and the convergence of λ_k to $\bar{\lambda}$ is rapid. However, the functions are assumed twice continuously differentiable.

Consider now the optimal control problem (OC) in 5.2. Assume that the hypotheses of Theorem 5.3.4 are fulfilled; so, in particular, the Lagrange multiplier $\tau = 1$. Assume also that there is some effective algorithm for solving uniquely the system consisting of the given differential equation $Dx = M(x, u)$, including boundary conditions on x, together with the adjoint differential equation, for the trajectory x and the Lagrange multiplier $\overline{\lambda}$, for each assumed control function u. (Such an algorithm is necessarily more complicated than one for a pair of finite difference equations, as in 4.6; some finite difference approximations to differential equations are numerically unstable, allowing errors to grow rapidly.) If the assumed control is u_n, denote these solutions by x_n and $\overline{\lambda}_n$. Assume that

$$H_n(u) \equiv \tau F(x_n, u) + \overline{\lambda}_n M(x_n, u) = \int_I h_n(u(t), t)\, \mathrm{d}t.$$

For the associated problem (OC#), assume that $H_n(u)$ is minimized, with respect to $G(u) \in K$, at $u = v_n$. Now set $u_{n+1} = \Theta_n(u_n, v_n)$, where Θ_n is some suitable function, and iterate again. If Θ_n is such that $F(x_{n+1}, u_{n+1})$ is less than $F(x_n, u_n)$ by a sufficient amount, and also that $\|u_{n+1} - u_n\|$ is sufficiently small, then every accumulation point of $\{u_n\}$ is an optimal control. This approach involves both a sequence $\{u_n\}$ of controls, and a sequence $\{\overline{\lambda}_n\}$ of Lagrange multipliers.

The choice of the function Θ_n is critical. Teo (1977) has obtained rapid convergence with $\Theta_n(u_n, v_n)(t) = v_n(t)$ for $t \in I_n$, and $\Theta_n(u_n, v_n)(t) = u_n(t)$ otherwise, where I_n is a suitably constructed subset of I. This approach also applied to a suitable partial differential equation $Dx = M(x, u)$. (See Mayne and Polak, 1975; Teo, Reid, and Boyd, 1977.)

7.6 Quadratic programming by Beale's method

Wolfe's method (see 4.6) for the quadratic program

Minimize $-c^T x + \frac{1}{2} x^T P x$ subject to $Ax = b, x \geqslant 0, x \in \mathbb{R}^n$,

where P is symmetric, has the serious disadvantage of requiring a very large matrix, including both A and A^T as submatrices, in the calculation. Beale (1967) has given an alternative method, which instead uses A, with a variable number of additional rows adjoined; published examples of Beale's method indicate tableaus considerably smaller than for Wolfe's method.

Consider first the simplex method, applied to a linear program with constraints $Ax = b$, $x \geqslant 0$. At a particular iteration, the basis is specified by a square submatrix $A^{(1)}$ of A. Denote the inverse of $A^{(1)}$ by Q, and partition $A = [A^{(1)} \mathrel{\vdots} A^{(2)}]$ and correspondingly $x^T = [y^T \mathrel{\vdots} z^T]$. The basic feasible solution is $y = Qb$, $z = 0$. If any of the nonbasic variables z are increased from zero, then $y = Qb - QA^{(2)}z$.

Suppose now that the objective function is quadratic, as given above; the matrix P is not necessarily positive semi-definite, but the program is assumed to have a finite minimum. Substituting for x in terms of z from

$$x = \begin{bmatrix} Qb - QA^{(2)}z \\ z \end{bmatrix} = \begin{bmatrix} Qb \\ 0 \end{bmatrix} - \begin{bmatrix} QA^{(2)} \\ -I \end{bmatrix} z \equiv q - Mz,$$

$f(x)$ is expressed in terms of z by

$$f(x) = F(z) = (- c^T q + \tfrac{1}{2} q^T P q)$$
$$+ (c - Pq)^T Mz + \tfrac{1}{2} z^T M^T PMz = h + d^T z + \tfrac{1}{2} z^T Gz,$$

where h, d, and G depend on the current basis. Then

$$\partial F(z)/\partial z_j = z^T G_j + d_j,$$

where G_j is column j of G, and d_j is element j of d.

Suppose, for some j, that $\partial F(z)/\partial z_j < 0$ for $0 < z_j < \delta_j$, and $= 0$ at $z_j = \delta_j$; then $\partial F(z)/\partial z_j > 0$ for $z_j > \delta_j$. Then $F(z)$ is decreased by increasing z_j from 0 until either (a) z_j reaches δ_j, or (b) some basic variable x_i is reduced to zero. If (a) comes first, then a new nonbasic variable $u = z^T G_j + d_j$ is introduced in place of z_j; note that u is a *free* variable, meaning that there is no requirement that $u \geqslant 0$, and that an extra equation has now been adjoined to $Ax = b$. The corresponding Q, thus expanded by a row and a column, can be calculated from the previous Q by a partitioned matrix inverse formula, namely

$$
\begin{bmatrix} A^{(1)} & \vdots & r \\ \cdots & \cdots & \cdots \\ s^T & \vdots & t \end{bmatrix}^{-1} = \begin{bmatrix} Q - Qrms^T Q & \vdots & -Qrm \\ \cdots\cdots\cdots\cdots\cdots & & \cdots\cdots \\ -ms^T Q & \vdots & m \end{bmatrix},
$$

where Q is the inverse of $A^{(-1)}$, and $m = (t - s^T Q r)^{-1}$. This procedure does not require a new matrix inverse to be calculated from the beginning.

If, instead, (b) comes first, an ordinary simplex iteration is done, replacing x_i in the basis by z_j. Suppose, after this, that the new $F(z)$ contains s free variables; let u_1 be one of them; then the terms involving u_1 have the form $\phi = u_1(\frac{1}{2}\lambda u_1 + g^T w)$, where w is the vector of remaining nonbasic variables. Here $\lambda > 0$, otherwise the objective function is unbounded below on its constraint set. Setting $v_1 = \partial\phi/\partial u_1 = \lambda u_1 + g^T w$ as a new free variable to replace u_1, a calculation similar to (a) transforms ϕ to the form

$$
(2\lambda)^{-1}(v_1^2 - (g^T w)^2).
$$

After at most s such transformations, $F(z)$ has been reduced to *standard form*, namely where $d_j = 0$ whenever z_j is a free variable.

If $F(z)$ is in standard form, and all nonbasic non-free z_j are equated to zero, then $F(z)$ is minimized, with respect to the free variables, when $F(z) = h$, its constant term. Hence h is a function only of the set E of nonbasic non-free variables. Assuming no degeneracy, h strictly decreases after each iteration, so no set E is encountered twice. There are only finitely many such sets E; and the numbers s are bounded above (by n). Hence a minimum to the quadratic programming problem is reached, by this method, in a finite number of iterations, just as the simplex method does for linear programming.

7.7 Decomposition

Consider the linear programming problem

(P1): Minimize $\sum_{j=1}^{r} c_j^T x_j$ subject to $Cx = b$, $(\forall j)x_j \geqslant 0$,

where

$$
x = \begin{bmatrix} x_1 \\ x_2 \\ . \\ . \\ . \\ x_r \end{bmatrix}, \quad
C = \begin{bmatrix} A_1 & A_2 & \ldots & A_r \\ B_1 & 0 & \ldots & 0 \\ 0 & B_2 & \ldots & 0 \\ . & . & \ldots & . \\ 0 & 0 & \ldots & B_r \end{bmatrix}, \quad
b = \begin{bmatrix} b_0 \\ b_1 \\ b_2 \\ . \\ b_r \end{bmatrix},
$$

where the dimensions of the matrices are $A_j(m_0 \times n_j)$, $B_j(m_j \times n_j)$, $b_j(m_j \times 1)$, x_j and c_j ($n_j \times 1$). Such a problem may arise in optimal planning over r successive time periods, where the constraints $B_j x_j = b_j$, $x_j \geq 0$ apply to period j, and in addition there are overall constraints $\Sigma_{j=1}^r A_j x_j = b_0$, which may represent, for example, limits on the total supply of raw material to a factory, over all the time periods.

Various *decomposition methods* exist, which solve (P1) by solving several linear programs involving smaller matrices than the large matrix C. This is computationally useful, notably when all the B_j are the same, as may well happen. Also the optimization of the subproblems – one for each B_j – in a semi-independent manner has economic significance. The methods discussed here are due to Dantzig and Wolfe (1960), and to Bennett (1966).

For the first method, assume, for each j, that the polyhedron $S_j = \{x_j \in \mathbb{R}^{n_j}_+ : B_j x_j = b_j\}$ is bounded. Then S_j has finitely many (say s_j) extreme points; let E_j be the matrix whose columns are the vectors of these extreme points. Then any $x_j \in S_j$ can be written as $x_j = E_j v_j$ where $v_j \geq 0$ and $e_{s_j}^T v_j = 1$, where e_k denotes a column of k ones. (Thus x_j is expressed as a convex combination of the extreme points.)

Then (P1) can be transformed to the equivalent problem

(P2): Minimize $\sum_{j=1}^r d_j^T v_j$ subject to $Mv = m$ and $(\forall j)v_j \geq 0$,

where

$$
M = \begin{bmatrix}
M_1 & M_2 & \ldots & M_r \\
e_{s_1}^T & 0 & \ldots & 0 \\
0 & e_{s_2}^T & \ldots & 0 \\
\cdot & \cdot & \ldots & \cdot \\
0 & 0 & \ldots & e_{s_r}^T
\end{bmatrix}, \quad m = \begin{bmatrix}
b_0 \\ 1 \\ 1 \\ \cdot \\ 1
\end{bmatrix};
$$

with $M_j = A_j E_j$ and $d_j^T = c_j^T E_j$ (resp. $m_0 \times s_0$ and $1 \times s_j$ matrices). Thus (P2) is a problem of similar form to (P1), but with a much simpler matrix, fewer constraints, but many more variables. But it will turn out that only those columns of M which enter bases need to be computed.

A basic feasible solution of (P2) is specified by a subset of $m_0 + r$ columns of M; denote by Q the inverse of the sub-matrix formed by these basis columns; let the cost vector d^T correspond to this basis. A new column of M, corresponding to column k of M_j (denote it by $M_{j(k)}$), may enter the basis provided that (see 3.3)

$$
0 < p^T M_{j(k)} + q_j - d_{j(k)} \quad \text{where} \quad d^T Q = [p^T \; \vdots \; q^T],
$$

where p has m_0 components, and q has r components. An equivalent requirement is

$$
0 < (p^T A_j - c_j)w + q_j, \quad \text{where} \quad w = E_{j(k)}.
$$

An iteration for (P2) requires therefore the solution of r subproblems (SPj) ($j = 1, 2, \ldots, r$); (SPj) is the linear program:

(SPj): $\underset{x_j \in \mathbb{R}^{n_j}}{\text{Minimize}} \ (c_j - p^T A_j)x_j$ subject to $x_j \geqslant 0, B_j x_j = b_j$.

Suppose that (SPj) attains its minimum t_j at $x_j = \bar{x}_j$. Then $-t_j + q_j$ is maximized over $j \in \{1, 2, \ldots, r\}$ at $j = \bar{j}$ and $x_{\bar{j}} = E_{\bar{j}(\bar{k})}$ for some $\bar{j}$ and $\bar{k}$. The new column to enter the basis of (P2) is then specified by $\bar{j}, \bar{k}$, and $E_{\bar{j}(\bar{k})}$; note that this is enough now to calculate the required column $M_{\bar{j}(\bar{k})}$.

Since this algorithm is equivalent to the simplex method applied directly to (P2), it converges in a finite number of iterations to an optimum of (P2), and therefore of (P1).

Denote by $\bar{p}, \bar{q}$ the vectors p, q for the final (optimal) iteration of (P2). For small enough perturbations of the requirements vector b about its given value, the optimal basis will not change. For such perturbations, the optimum of (P1) is given by solving the r independent subproblems (SPj), with p fixed at $\bar{p}$. Over this range of perturbations, the subsystems represented by $\{B_j, b_j\}$ may be solved independently, and the overall optimum obtained, provided that the cost vectors c_j are modified to $c_j - \bar{p}^T A_j$ $(j = 1, 2, \ldots, r)$. This provides a model for decentralized management of an interconnected economic system.

Bennett's alternative approach to solving (P1) considers a basic feasible solution for (P1), and reorders the basis columns to the form

$$
\begin{bmatrix}
A_1^{(2)} & A_2^{(2)} & \cdots & A_r^{(2)} & A_1^{(3)} & A_1^{(3)} & \cdots & A_r^{(3)} \\
B_1^{(2)} & 0 & \cdots & 0 & B_1^{(3)} & 0 & \cdots & 0 \\
0 & B_2^{(2)} & \cdots & 0 & 0 & B_2^{(3)} & \cdots & 0 \\
\cdot & \cdot & \cdots & \cdot & \cdot & \cdot & \cdots & \cdot \\
0 & 0 & \cdots & B_r^{(2)} & 0 & 0 & \cdots & B_r^{(3)}
\end{bmatrix} \equiv W.
$$

Here $\begin{bmatrix} A_j^{(2)} \\ B_j^{(2)} \end{bmatrix}$ and $\begin{bmatrix} A_j^{(3)} \\ B_j^{(3)} \end{bmatrix}$ are submatrices of $\begin{bmatrix} A_j \\ B_j \end{bmatrix}$, chosen so that each $B_j^{(2)}$ is square and nonsingular. Bennett then applies formulas for partitioned matrix inverse to update the inverse of W from each iteration to the next. (Bennett's notation has been changed, and the order of the rows altered, to agree with (P1) above. This does not affect the principle of the method. Bennett also uses superscript (1) for nonbasic variables, which are not required here.)

If a *nonlinear* programming problem can be adequately approximated by a separable programming problem (3.4), it is then possible to apply decomposition methods to the resulting linear program. However, the restriction on the number of basic variables in any special ordered set may be fulfilled in every subproblem, and yet violated in the problem as a whole. It would be necessary to check, at the end, whether such a violation had occurred.

Rosen (1963), and Rosen and Ornea (1963), have given an extension of decomposition methods to problems where the equations linking the subproblems are nonlinear. (In the linear case, the linking equations are those involving the A_j.) The subproblems are linear, but there remains a nonlinear problem, which Rosen solves by gradient projection, in each iteration of the main problem.

References

Abadie, J. (Ed.) (1967), *Nonlinear Programming*, North-Holland, Amsterdam.

Abadie, J. (Ed.) (1970), *Integer and Nonlinear Programming*, North-Holland, Amsterdam.

Beale, E.M.L. (1967), Numerical methods, Chapter 7 of Abadie (1967).

Beale, E.M.L. (1970), Advanced algorithmic features for general mathematical programming systems, Chapter 4 of Abadie (1970).

Bennett, J.M. (1966), An approach to some structured linear programming problems, *Operations Research*, **14**, 636–645.

Box, M.J., Davies, D., and Swann, W.H. (1969), *Non-linear Optimization Techniques*, Oliver and Boyd, Edinburgh. (I.C.I. Mathematical and Statistical Techniques for Industry, Monograph No. 5.)

Broyden, C.G. (1972), Quasi-Newton methods, Chapter 6 of Murray (1972).

Dantzig, G.B., and Wolfe, P. (1960), A decomposition principle for linear programs, *Operations Research*, **8**, 101–111.

Fiacco, A.V., and McCormick, G.P. (1968), *Nonlinear Programming: Sequential Unconstrained Minimization Techniques*, Wiley, New York.

Gill, P.E., and Murray, W. (Eds.) (1974), *Numerical Methods for Constrained Optimization*, Academic Press, London.

Lootsma, F.A. (1969), Hessian matrices of penalty functions for solving constrained minimization problems, *Philips Res. Reports*, **24**, 332–330.

Luenberger, D.G. (1969), *Optimization by Vector Space Methods*, Wiley, New York.

Luenberger, D.G. (1973), *Introduction to Linear and Nonlinear Programming*, Addison-Wesley, Reading.

Mangasarian, O.L. (1974), *Nonlinear Programming, Theory and Computation*, Chapter 6 of Elmaghraby, S.E., and Moder, J.J. (Eds.),

Handbook of Operations Research, Van Nostrand Reinhold, New York.

Mayne, D.Q., and Polak, E. (1975), First-order strong variation algorithms for optimal control, *J. Optim. Theor. Appl.*, **16**, 277–301.

Miele, A. (1975), Recent advances in gradient algorithms for optimal control problems, *J. Optim. Theor. Appl.*, **17**, 361–430.

McCormick, G.P., and Pearson, J.D. (1969), Variable metric methods and unconstrained optimization, in Fletcher, R. (Ed.), *Optimization*, Academic Press, London and New York.

Murray, W. (ed.) (1972), *Numerical Methods for Unconstrained Optimization*, Academic Press, London.

Pietrzykowski, T. (1969), An exact potential method for constrained maximum, *SIAM J. Numer. Anal.*, **6**, 299–304.

Polak, E. (1971), *Computational Methods in Optimization*, Academic Press, New York.

Rockafellar, R.T. (1974), Augmented Lagrange multiplier functions and duality in non-convex programming, *SIAM J. Control*, **12**, 268–287.

Rosen, J.B. (1963), in Graves, R.L., and Wolfe, P., *Recent Advances in Mathematical Programming*, McGraw-Hill, New York, 159–176.

Rosen, J.B., and Ornes, J.C. (1963), Solution of nonlinear programming problems by partitioning, *Management Science*, **10**, 160–173.

Teo, K.-L. (1977), Computational methods of optimal control problems with application to decision making of a business firm, University of New South Wales, Private communication.

Teo, K.-L., Reid, D.W., and Boyd, I.E. (1977), Stochastic optimal control theory and its computational methods, University of New South Wales, Research Report.

Wolfe, P. (1967), Chapter 6 of Beale (1967).

Appendices

A.1 Local solvability

Let X and Y be Banach spaces, X_0 a convex open subset of X, $a \in X_0$, and $g : X_0 \rightarrow Y$ a continuously Fréchet differentiable function. Then the Fréchet derivative $g'(u)$ is a continuous function of u, so that $\|g'(x + z) - g'(a)\| < \epsilon$ whenever $\|x - a\| < \delta(\epsilon)$ and $\|z\| < \delta(\epsilon)$, for some function $\delta(\cdot)$, and also

$$g(x + y) - g(x) = \int_0^1 \left[\frac{\partial}{\partial \alpha} g(x + \alpha y) \right] d\alpha = \int_0^1 g'(x + \alpha y) y \, d\alpha.$$

Therefore, whenever $\|x - a\| < \delta(\epsilon)$ and $\|y\| < \delta(\epsilon)$,

$$g(x + y) - g(x) = g'(a)y + \xi(x, y),$$

where

$$\|\xi(x, y)\| = \left\| \int_0^1 [g'(x + \alpha y) - g'(a)] y \, d\alpha \right\|$$

$$< \int_0^1 \epsilon \|y\| \, d\alpha = \epsilon \|y\|.$$

Implicit function theorem. Let X and Y be Banach spaces, X_0 a convex open subset of X, and $g : X_0 \rightarrow Y$ a continuously Fréchet differentiable function; at $a \in X_0$, let $g(a) = 0$ and

147

let $g'(a)(X) = Y$. Then, whenever the direction c satisfies $g'(a)c = 0$, there exists a solution $x = a + \alpha c + \eta(\alpha)$ to $g(x) = 0$, valid for sufficiently small $\alpha > 0$, where $\|\eta(\alpha)\|/\alpha \to \alpha$ as $\alpha \downarrow 0$.

Remarks. If $g'(a)(X) = Y$, then $g(x) = 0$ is then *locally solvable* (see 2.6) at a. The solution for x may be written $x = a + \alpha c + o(\alpha)$. To obtain the form of the implicit function theorem required in 4.8, consider the equation $h(x,y) = 0$, where h is continuously Fréchet differentiable; x, y, and $h(x, y)$ take values in Banach spaces; $h(0, 0) = 0$; and $A = h_x(0, 0)$ is assumed invertible. Denote $B = h_y(0, 0)$. If the vector (p, q) satisfies $Ap + Bq = 0$, the above theorem finds a solution $x = \alpha p + o(\alpha)$, $y = \alpha q + o(\alpha)$ to $h(x, y) = 0$, valid for sufficiently small $\alpha > 0$. If $q \neq 0$, then each $o(\alpha) = o(\|y\|)$; and A^{-1} exists; hence

$$x = -A^{-1}B(\alpha q) + o(\alpha) = -A^{-1}By + o(\|y\|)$$

satisfies $h(x, y) = 0$, whenever $\|y\|$ is sufficiently small.

Proof. Since $M = g'(a)$ is a continuous linear map of X onto Y, the open mapping theorem states that $M(\{x : \|x\| < 1\})$ contains some open ball $\{y : \|y\| < \lambda\}$ in Y; fix any $\rho > \lambda^{-1}$. From this and $M(X) = Y$ it follows that to any $y \in Y$ there corresponds (not usually uniquely) an $x = \phi(y) \in X$ such that $M\phi(y) = -y$ and $\|\phi(y)\| < \rho\|y\|$.

Let c satisfy $Mc = 0$, $0 < \|c\| \leqslant 1$. Let $\epsilon > 0$ satisfy $\epsilon\rho < \frac{1}{2}$; find $\delta(\epsilon)$ from g as in the first paragraph of A.1; let α satisfy $0 < \alpha < \delta(\epsilon)/2$. Construct the sequence $\{x_n\} \subset X$ by $x_0 = a + \alpha c$, and

$$x_n = x_{n-1} + \phi \circ g(x_{n-1}) \quad (n = 1, 2, \dots).$$

Then $x_1 - x_0 = \phi \circ (\alpha Mc + \xi(a, \alpha c)) = \phi \circ \xi(a, \alpha c)$, so that $\|x_1 - x_0\| < \rho \|\xi(a, \alpha c)\| < \rho\epsilon\alpha$. Suppose, for $j \leqslant n - 1$ and $n > 1$, that $\|x_j - x_{j-1}\| < \alpha(\rho\epsilon)^j$. Then

$$x_n - x_{n-1} = \phi \circ (g(x_{n-2}) + M(x_{n-1} - x_{n-2})$$
$$+ \xi(x_{n-2}, x_{n-1} - x_{n-2})) = \phi \circ \xi(x_{n-2}, x_{n-1} - x_{n-2}),$$

since $M(x_{n-1} - x_{n-2}) = -g(x_{n-2})$. Also

$$\|x_{n-2} - a\| < \alpha\|c\| + \sum_{j=1}^{\infty} \alpha(\rho\epsilon)^j$$

$$\leqslant \alpha + \alpha\rho\epsilon/(1 - \rho\epsilon) < \alpha + \alpha < \delta(\epsilon)$$

since $\rho\epsilon < \frac{1}{2}$. Hence $\|x_n - x_{n-1}\| < \rho\epsilon[\alpha(\rho\epsilon)^j]$. Hence, by induction, $\|x_j - x_{j-1}\| < \alpha(\rho\epsilon)^j$ holds for all j. Since X is complete, the sequence $\{x_n\}$ converges, by comparison with the geometric series $\Sigma\alpha(\rho\epsilon)^j$, to some $\bar{x} \in X$; since $\|\bar{x} - a\| \leqslant \delta(\epsilon)$, $\bar{x} \in X_0$. Since M is continuous, $g(\bar{x}) = \lim g(x_{n-1}) = \lim M(x_{n-1} - x_n)$, thus $g(\bar{x}) = 0$. And $\bar{x} = a + \alpha c + \eta$ where

$$\|\eta\| = \|\bar{x} - x_0\| \leqslant \sum_{j=1}^{\infty} \alpha(\rho\epsilon)^j < \alpha\rho\epsilon/(1 - \rho\epsilon) < 2\alpha\rho\epsilon.$$

Extensions. Under the stronger hypothesis that $g'(a)$ has *full rank* (1.8), the map ϕ may be assumed linear, and then the contraction mapping theorem can be used to show that the map ψ of αc to $a + \alpha c + \eta(\alpha)$ takes a neighbourhood of 0 in the subspace $g'(a)^{-1}(0)$ homeomorphically *onto* a neighbourhood of a in the manifold $g^{-1}(0)$, and that $\psi'(0)$ is the identity map. (See Craven, 1970; Flett, 1966).

Consider now the system $g(x) \in -S$, where $S \subset Y$ is a closed convex cone, and $g(a) = b \in -S$; denote $M = g'(a)$. Now let c satisfy $0 < \|c\| \leqslant 1$ and $b + Mc \in -S$, thus $Mc = -q - b$ for some $q \in S$. Then, for $0 < \alpha < 1, -s_0 \equiv b + \alpha Mc = (1 - \alpha)b - \alpha q \in -S$. Define the sequence $\{x_n\}$ by $x_0 = a + \alpha c$ and, *assuming* $M(X) = Y$, $x_n = x_{n-1} + \phi(g(x_{n-1}) + s_0)$ for $n = 1, 2, \ldots$, with ϕ as in the preceding proof. Then

$$\|g(x_0) + s_0\| = \|s_0 + b + \alpha Mc + \xi(a, \alpha c)\| < \epsilon\|\alpha c\| \leqslant \epsilon\alpha.$$

The proof of the above theorem, with $g(x_n) + s_0$ replacing $g(x_n)$, then shows that $\{x_n\}$ converges to $\bar{x} \in X_0$, where $\bar{x} = a + \alpha c + o(\alpha)$ as $\alpha \downarrow 0$ and $g(\bar{x}) + s_0 = 0$, thus $g(\bar{x}) \in -S$. Thus local solvability of $g(x) \in -S$ is proved. However, the hypothesis $M(X) = Y$ assumes too much.

Theorem (Local Solvability). Let X and Y be Banach spaces, X_0 a convex open subset of X, $S \subset Y$ a closed convex cone, and $g: X_0 \to Y$ a continuously Fréchet differentiable function. At $a \in X_0$, let $b = g(a) \in -S$, and let $b + M(X) + S \supset N$, where $M = g'(a)$ and N is some ball with centre 0 in Y. Then $-g(x) \in S$ is locally solvable at $a \in X_0$.

Proof. Let nonzero c satisfy $b + Mc \in -S$; then (as above) $b + \alpha Mc \in -S$ if $0 < \alpha < 1$. The function $f(\alpha, x) = g(x + \alpha c)$ is continuously Fréchet differentiable in (α, x). Then, given the stated hypotheses on b, M, N, S, a theorem of S. Robinson (stated below) shows that, whenever α is sufficiently small, $f(\alpha, x) \in -S$ has a solution $x = a + u(\alpha)$, and, for some constant γ, the distance from a to the solution set of $f(\alpha, x) \in -S$ is less than $\gamma r(\alpha)$, where $r(\alpha) = d(0, f(\alpha, a) + S)$ and d denotes distance. Consequently $\|u(\alpha)\| \leqslant 2\gamma r(\alpha)$. From $g(a + \alpha c) = b + \alpha Mc + o(\alpha)$ and $b + \alpha Mc \in -S$, it follows that $r(\alpha) = o(\alpha)$. Hence $u(\alpha) = o(\alpha)$ as $\alpha \downarrow 0$, and $g(a + \alpha c + u(\alpha)) = f(\alpha, a + u(\alpha)) \in -S$.

Expressed in the present notation, Robinson's theorem (see Robinson, 1974, Corollary 1) states, in part, the following. Let X and Y be Banach spaces, X_0 an open subset of X, P a topological space, $S \subset Y$ a closed convex cone, $C \subset X_0$ a closed convex set, $f: P \times X_0 \to Y$ a function Fréchet differentiable with respect to x on $P \times X_0$, with f and its derivative f_x continuous in (α, x) at $(0, a)$. Denote $g(x) = f(0, x)$. Assume that

$$0 \in \mathrm{int}[g(a) + g'(a)(-a + C) + S] \quad \text{and} \quad g(a) \in -S, a \in C.$$

Then $f(\alpha, x) \in -S$ has a solution $x \in C$ whenever α is near enough to 0 in P, and the distance from a to the solution set of $f(\alpha, x) \in -S$ is less than $\gamma d(0, f(\alpha, a) + S)$, for some constant γ. For the local solvability theorem, $C = X$ is substituted.

Remark. Consider a pair of constraints $-g(x) \in S$ and $-h(x) \in T$, where $g: X_0 \to Y$ and $h: X_0 \to Z$ are continuously Fréchet differentiable, and $S \subset Y$ and $T \subset Z$ are closed convex cones. Suppose that

$$g(a) + g'(a)(X) + S \supset N_1 \quad \text{and} \quad h(a) + h'(a)(X) + T \supset N_2,$$

where $N_1 \subset Y$ and $N_2 \subset Z$ are balls with centre 0. Then $N = N_1 \times N_2$ is a ball with centre 0 in $Y \times Z$, with the norm $\|y\| + \|z\|$, and hence the pair of constraints

$$-\begin{bmatrix} g(x) \\ h(x) \end{bmatrix} \in \begin{bmatrix} S \\ T \end{bmatrix} \quad \text{satisfies} \quad \begin{bmatrix} g(a) \\ h(a) \end{bmatrix} + \begin{bmatrix} g'(a) \\ h'(a) \end{bmatrix}(X) + \begin{bmatrix} S \\ T \end{bmatrix} \supset N.$$

Hence the pair of constraints is locally solvable, if each constraint satisfies the solvability criterion of the above theorem.

References

Craven, B.D. (1970), A generalization of Lagrange multipliers, *Bull. Austral. Math. Soc.*, 3, 353–362. (See Lemma 3.)

Flett, T.M. (1966), *Modern Analysis*, McGraw-Hill, New York.

Robinson, Stephen M. (1974, 1976), Stability theory for systems of inequalities; Part II: Differentiable nonlinear systems, University of Wisconsin, Mathematics Research Center, Technical Summary Report No. 1388. And *SIAM J. Numerical Analysis*, 13, 497–513.

A.2 On separation and Farkas theorems

The separation theorem quoted in 2.2.3 may be found, except for the last statement, in Schaefer (1966). Weak $*$ neighbourhoods in X' are defined in 1.8; weak $*$ neighbourhoods in X are similarly defined, but with X and X' interchanged. If continuity of linear functionals is defined in terms of weak neighbourhoods in X and weak $*$ neighbourhoods in X', then the dual of X' becomes X. Since X and X' are now related symmetrically, the separation theorem can be used to find an $x \in X$ separating a convex set $K \subset X'$ and a point $b \in X' \backslash K$, provided that K is closed in the appropriate, weak $*$, sense. This gives the last statement of 2.2.3.

If S is a convex cone in X then S^* is weak $*$ closed in X', since S^* is the intersection, over $x \in S$, of weak $*$ closed sets $\{y \in X^* : yx \geqslant 0\}$.

In 2.2.7, a linear map h is constructed; it remains to prove that h is continuous. Let p denote the projection of Y onto the quotient space $Y/C^{-1}(0)$; define a set $U \subset p(Y)$ to be open iff $p^{-1}(U)$ is open in Y. Since $c = hC$, $cp^{-1} = h(Cp^{-1})$, hence $h = (cp^{-1})(Cp^{-1})^{-1}$; here Cp^{-1} is single valued by construction, and cp^{-1} is single valued since $Cy = 0$ implies $cy = 0$. Then h is continuous provided that cp^{-1} and $(Cp^{-1})^{-1}$ are continuous.

Let U be any open set in $p(Y)$. Since $C = (Cp^{-1})p$, and C is continuous, $(Cp^{-1})^{-1}(U) = p(C^{-1}(U))$ is open, hence Cp^{-1} is continuous. A similar argument shows that cp^{-1} is continuous. Now $U = p(V)$ for some open $V \subset Y$, hence also $(Cp^{-1})(U) = C(V)$; but $C(Y) = X$, so the open mapping theorem shows that C maps open sets to open sets; since also Cp^{-1} is bijective, it follows that $(Cp^{-1})^{-1}$ is defined and continuous.

References

Craven, B.D. (1972), Nonlinear programming in locally convex spaces, *J. Optim. Theor. Appl.*, **10**, 197–210. (See Theorems 2.1 and 2.2.)

Craven, B.D., and Koliha, J.J. (1977), Generalizations of Farkas's theorem, *SIAM J. Math. Anal.*, **8**, 983–997.

Schaefer, H.H. (1966), *Topological Vector Spaces*, Macmillan, New York.

A.3 A zero as a differentiable function

For an interval $I \subset \mathbb{R}$, denote by $C^1(I)$ the space of continuously differentiable real functions on I, with the norm $\|f\| = \|f\|_\infty + \|Df\|_\infty$, where D denotes derivative, and $\|\cdot\|_\infty$ is the uniform norm on I. Assume $0 \in \text{int } I$.

Theorem. Let $f \in C^1(I)$ have a zero at 0, such that $f'(0) \neq 0$. Then, for sufficiently small $\|g\|$, $f + g \in C^1(I)$ has a zero at $x(g)$, where $x(\cdot)$ is a Fréchet differentiable function of $g \in C^1(I)$ at 0.

Proof. Let $t, u \in I$, and $0 < \lambda \leqslant \frac{1}{4}|f'(0)|$. Since $f \in C^1(I)$ and $f(0) = 0$, $f(t) = f'(0)t + \theta(t)$ where, for each λ, $|\theta(t) - \theta(u)| < \lambda|t - u|$ whenever $|t| < \delta(\lambda)$, $|u| < \delta(\lambda)$, using A.1 (first paragraph). Also,

$$|g(t) - g(u)| = \left| \int_0^1 g'(u + \alpha(t - u))\, d\alpha \right| \leqslant \|g\| \, |t - u|.$$

Now $(f + g)(t) = f'(0)t + \theta(t) + g(t) = 0$ iff $t = H(t)$, where, noting that $f'(0) \neq 0$, $H(t) = -[f'(0)]^{-1}[\theta(t) + g(t)]$. Combining these results, whenever $|t| < \delta(\lambda)$, $|u| < \delta(\lambda)$, and

$$\|g\| < \min\{\lambda, \tfrac{1}{2}\delta(\lambda)|f'(0)|\},$$

$$|H(t) - H(u)| \leqslant |f'(0)|^{-1}(\lambda + \|g\|)|t - u| \leqslant \epsilon|t - u|,$$

where $\epsilon = 2\lambda/|f'(0)| \leqslant \frac{1}{2}$, and $|H(0)| = |f'(0)|^{-1}|g(0)| \leqslant \frac{1}{2}\delta(\lambda)$.

Define a sequence $\{t_n\}$ by $t_0 = 0$, and $t_{n+1} = H(t_n)$ for $n = 0, 1, 2, \ldots$. If $|t_{j+1} - t_j| < \epsilon^j|t_1|$ whenever $j \leqslant n - 1$, then

$$|t_{n+1}| \leqslant \sum_{j=0}^{n} |t_{j+1} - t_j| < \sum_{j=0}^{n} \epsilon^j|t_1| \leqslant 2|t_1| < \delta(\lambda);$$

and then $|t_{n+1} - t_n| \leqslant \epsilon|t_n - t_{n-1}| < \epsilon(\epsilon^n|t_1|) = \epsilon^{n+1}|t_1|$. Since $|t_1 - t_0| = |t_1| \leqslant \epsilon^0|t_1|$, $|t_{j+1} - t_j| < \epsilon^j|t_1|$ holds by induction for all j. Hence the series $\sum_{j=0}^{\infty}[t_{j+1} - t_j]$ converges, by comparison with the geometric series $\sum_{j=0}^{\infty} \epsilon^j|t_1|$, to a limit $x(g)$ satisfying $H(x(g)) = x(g)$. Then $x(g)$ is a zero of $f + g$; $t_1 = -[f'(0)]^{-1}g(0)$ is linear in g; and

$$|x(g) - t_1| \leqslant \sum_{j=1}^{\infty} |t_{j+1} - t_j| \leqslant \sum_{j=1}^{\infty} \epsilon^j t_1 \leqslant 2\epsilon|t_1|;$$

so $x(g)$ is a Fréchet differentiable function of $g \in C^1(I)$ at 0.

Extension. The conclusion remains valid if I is replaced by a closed ball in a Banach space X, and the functions f and g take values in a normed space Y, provided that the functions f and g are continuously Fréchet differentiable, and the hypothesis $f'(0) \neq 0$ is replaced by the hypothesis that $f'(0)$ has full rank (as defined in 1.8). (If $X = \mathbb{R}^n$ and $Y = \mathbb{R}^m$, then full rank requires that $m \leqslant n$.) Let $M = f'(0)$; define M_1 from M as in 1.8. In the above proof, $[f'(0)]^{-1}$ is then replaced by M_1^{-1}, and $|\cdot|$ by $\|\cdot\|$.

A.4 Lagrangean conditions when the cone has empty interior

In the Fritz–John theorem 4.4.1, one constraint $- h(x) \in T$ must satisfy a local solvability and a closed-cone condition, the other constraint $- g(x) \in S$ must satisfy int $S \neq \emptyset$. The following counter example shows that these hypotheses cannot both be omitted. Let $I = [0, 1]$, $X = L^1(I)$; then each vector in X' is represented by a bounded function. Let Q be the convex cone of nonnegative functions in $L^1(I)$; it is readily shown that int $Q = \emptyset$. Define the continuous linear map $M : X \to X$ by $x(t) \to m(t)x(t) (t \in I)$, where $m(t) = t^{-1}$ on $(0, 1]$, $m(0) = 1$; then $Mx \in Q$ iff $x \in Q$. Hence $f(x) = \int_I x(t) \, dt$ is minimized, subject to $Mx \in Q$, at $x = 0$. If (FJ) holds, then $\tau = \lambda(t)m(t)$ for almost all $t \in I$, where $\lambda(\cdot)$ is a bounded function representing a Lagrange multiplier in Q^*, and λ and $\tau \geqslant 0$ are not both zero. But, since $m(t) = t^{-1}$ and $\lambda(t)$ is bounded, $\tau = 0$; then $\lambda(t) = t\tau = 0$ also. So (FJ) does not hold. (See also Craven, 1977.)

Suppose now that $S^* \subset Y'$ has a set E of *generators*, thus that $S^* = \{\alpha y : \alpha \in \mathbb{R}_+, y \in B\}$ where $0 \notin B = \overline{\text{co}}\ E$, for which B is weak $*$ compact. Define a linear map Φ of Y into $C(B)$, the space of continuous real functions on B with the uniform norm, by $(\forall y \in Y, \forall b \in B) (\Phi y)(b) = b(y)$. Let K be the cone $\{\psi \in C(B) : \psi(B) \subset \mathbb{R}_+\}$. Then int $K \neq \emptyset$; Φ is continuous, giving B the weak $*$ topology; and, using the separation theorem (2.2.3), $s \in S$ iff $(\forall b \in B) \, bs \geqslant 0$ iff $\Phi s \in K$. Hence the constraint $- g(x) \in S$ can be replaced by the equivalent constraint $- (\Phi g)(x) \in K$, with int $K \neq \emptyset$; and, since g is assumed Fréchet (or Hadamard) differentiable, so is Φg, with $(\Phi g)'(a) = \Phi[g'(a)]$. So Theorem 4.4.1 can be applied with the constraint $- (\Phi g)(x) \in K$.

Now (FJ) includes a term $\lambda(\Phi g)'(a)$, with $\lambda \in K^*$, and an equation $\lambda(\Phi g)(a) = 0$. Now $\lambda(\Phi g)'(a) = \lambda\Phi g'(a) = vg'(a)$,

where Φ^T is the transpose of Φ (see 1.8) and $v = \Phi^T\lambda \in Y'$.
Similarly $\lambda(\Phi g)(a) = 0$ gives $vg(a) = 0$. For each $s \in S$,
$vs = \lambda(\Phi s) \geqslant 0$ since $\Phi s \in K$; so $v \in S^*$. Since B is compact,
the Riesz representation theorem represents $\lambda \in C(B)'$ by a
signed measure μ, so that $\lambda\psi = \int_B \mu(db)\psi(b)$ for each
$\psi \in C(B)$; since $\lambda \in K^*$, μ is a non-negative measure. If $v = 0$
then, for each $s \in S$, $0 = (\Phi^T\lambda)s = \int_B \mu(db)bs$, which is
contradicted, if $\mu \neq 0$, by choice of s to make $bs > 0$ for
suitable b; this is possible since $0 \notin B$. Hence $v = 0$ implies
$\mu(B) = 0$, hence $\lambda = 0$. So $\lambda \neq 0$ implies $v \neq 0$. Hence (FJ)
holds, with τ and v not both zero.

The set E exists, in particular, when there exists $h \in \text{int}\, S$.
For then $wh > 0$ for each nonzero $w \in S^*$, by 2.4.10. Define
$B = \{w \in S^* : wh = 1\}$; then $0 \notin B$; B generates S^* as required;
B is convex and weak $*$ closed. If $b \in B$ then $bh = 1$, and
$h + N \subset S$ for some ball N with centre 0, so $b(h + N) \subset \mathbb{R}_+$.
So, for each $n \in N$, $bn \geqslant -1$ and $b(-n) \geqslant -1$, so
$\{\|b\| : b \in B\}$ is bounded. The Banach–Steinhaus theorem
then shows that the weak $*$ closed set B is weak $*$ compact;
the Krein–Milman theorem then shows that $B = \overline{\text{co}}\, E$, where
$E = \text{extr}\, B$ (see 2.1, and Schaefer (1966), cited in A.2).

Reference

Craven, B.D. (1977), Lagrangean conditions and quasiduality, *Bull.
Austral. Math. Soc.*, **16**, 325–339.

A.5 On measurable functions

Theorem 5.3.1 extends as follows when ϕ is not continuous,
but measurable. A point $t_0 \in I$ is a *point of density* of a set
$E \subset I$ if $\lim_{\tau \downarrow 0}(2\tau)^{-1}[m(E \cap [t_0 - \tau, t_0 + \tau]) = 1$, where m
denotes Lebesgue measure (thus $mE = \int_E dt$). A function
$k : I \to \mathbb{R}$ is *approximately continuous* at t_0 if there is a
measurable set $E \subset I$ such that t_0 is a point of density of E
(hence $m(E) > 0$) and

$$\lim_{t \to t_0, t \in E} k(t) = k(t_0).$$

A theorem (see Munroe, 1953, Section 42) then states that a measurable function k is approximately continuous almost everywhere in I; and the points of a measurable set $E \subset I$, with $m(E) > 0$, are almost everywhere points of density of E. From ϕ in Theorem 5.3.1, define $k(t) = \phi(t)$ for $t \in A^\#$, $k(t) = 0$ otherwise. Then k is approximately continuous almost everywhere in $A^\#$. Obtain $A_0^\#$ by excluding from $A^\#$ the sets of zero measure on which k is not approximately continuous, and which are not points of density of $A^\#$. Then any $t_0 \in A_0^\#$ is a point of density of $A^\#$, and there is a set $D \subset A^\#$, for which $m(D) > 0$ and $\lim_{t \to t_0, t \in D} k(t) = k(t_0)$, by definition of t_0. This limit shows that, by suitably restricting D to A, $\phi(t) = k(t) \geqslant \sigma$ for $t \in A$, for some $\sigma > 0$, and $m(A) > 0$. Then A_β is defined from A, as in Theorem 5.3.1.

Reference

Munroe, M.E. (1953), *Introduction to Measure and Integration*, Addison-Wesley, Reading.

A.6 Lagrangean theorems with weaker derivatives

The (FJ) and (KT) theorems of 4.4 remain true with a weaker definition of derivative than Fréchet, and this is needed for optimal control problems using other norms than the uniform norm.

Let $g : X \to Y$ have a *linear Gâteaux derivative* $g'(a)$ at $a \in X$, and let g satisfy the *Lipschitz condition*

$$(\forall x_1, x_2 \in X)\ \|g(x_1) - g(x_2)\| \leqslant k\|x_1 - x_2\|,$$

where the constant k does not depend on x_1 and x_2. Then for each arc $x = \omega(\alpha)$ in X, where ω is a continuous function of $\alpha \geqslant 0$ with $\omega(0) = a$, and having initial slope $s = \omega'(0)$,

$$\|g(\omega(\alpha)) - g(a + \alpha s)\| \leqslant k\|\omega(\alpha) - \omega(0) - \alpha\omega'(0)\| = o(\alpha)$$

as $\alpha \downarrow 0$, hence

$$\|g(\omega(\alpha)) - g(a) - \alpha g'(a)s\| \leqslant \|g(\omega(\alpha)) - g(a + \alpha s)\|$$

$$+ \|g(a + \alpha s) - g(a) - \alpha g'(a)s\|$$

$$= o(\alpha) + o(\alpha) = o(\alpha).$$

A function g with the latter property, that for each arc $\omega(\alpha)$ with $\omega(0) = a$ and having initial slope $\omega'(0)$,

$$g(\omega(\alpha)) - g(a) - \alpha g'(a)\omega'(0) = o(\alpha) \text{ as } \alpha \downarrow 0,$$

is called *Hadamard differentiable*.

The linearization theorem (2.6.1) assumes that f and g are Fréchet differentiable. However, the proof remains valid if f and g are only Hadamard differentiable, since this ensures that $\theta(\alpha) = o(\alpha)$ and $\rho(\alpha) = o(\alpha)$. Therefore the (FJ) theorem 4.4.1 for (P1) remains valid with f and g only linearly Gâteaux differentiable and Lipschitz, instead of Fréchet differentiable.

Example. Let $I \subseteq \mathbb{R}$ be an interval, $L^1(I)$ the Banach space of integrable real functions on I, with the norm $\|u\|_1 = \int_I |u(t)| \, dt$, $g : \mathbb{R} \times I \to \mathbb{R}$ a continuously differentiable function, such that $g(u, t)$ is Lipschitz in the variable u, and $\int_I |g(0, t)| \, dt$ finite. Denoting the Lipschitz constant by k,

$$\int_I |g(u(t), t)| \, dt \leq \int_I |g(0, t)| \, dt + \int_I k|u(t) - 0| \, dt < \infty;$$

hence g defines a function $G : L^1(I) \to L^1(I)$ by

$$(Gu)(t) = g(u(t), t) \quad (u \in L^1(I), t \in I).$$

This function G is linearly Gâteaux differentiable, with $G'(u)z(t) = g_u(u(t), t)z(t)$, since for each $t \in I$ and $u, z \in L^1(I)$,

$$k|z(t)| \geq |\alpha^{-1}[g(u(t) + \alpha z(t), t) - g(u(t), t)]|$$

$$= |\alpha^{-1}(G(u + \alpha z) - G(u))(t)|$$

$$= |g_u(u(t), t)z(t) + o(\alpha)/\alpha| \text{ as } \alpha \downarrow 0.$$

Hence

$$\|G'(u)z\|_1 = \int_I |g_u(u(t), t)z(t)| \, dt \leq k \int_I |z(t)| \, dt = k\|z\|_1.$$

Also, for any $u_1, u_2 \in L^1(I)$,

$$\| G(u_1) - G(u_2) \|_1 = \int_I |g(u_1(t), t) - g(u_2(t), t)| \, dt$$

$$\leqslant k \| u_1 - u_2 \|_1.$$

Hence G is Hadamard differentiable.

So also is the function ϕ, defined for $u \in L^1(I)$ by $\phi(u) = \int_I g(u(t), t) \, dt$, since $f \to \int_I f(t) \, dt$ is linear, and the chain rule is readily verified. Note that these calculations extend to u taking values in $\mathbb{R}^r$ and g taking values in $\mathbb{R}^h$, and $|\cdot|$ the Euclidean norm.

The application of this theory to the control problem of 1.10, with all uniform norms replaced by $L^1(I)$ norms, is discussed in 5.3.8.

A.7 On convex functions

Let $f: X \to Y$ be S-convex, where X and Y are normed spaces, and $S \subset Y$ is a closed convex cone. For $0 < \mu < \lambda$ and $a \in X$, $0 \neq y \in X$,

$$\mu \left\{ -\frac{f(a + \mu y) - f(a)}{\mu} + \frac{f(a + \lambda y) - f(a)}{\lambda} \right\}$$

$$= -f\left(\frac{\mu}{\lambda}(a + \lambda y) + \left(1 - \frac{\mu}{\lambda}\right)a \right) + \frac{\mu}{\lambda} f(a + \lambda y)$$

$$+ \left(1 - \frac{\mu}{\lambda}\right) f(a) \in S.$$

As $\mu \downarrow 0$, supposing that f is Fréchet differentiable,

$$-f'(a)y + [f(a + \lambda y) - f(a)]/\lambda \in S \quad \text{whenever} \quad \lambda > 0.$$

Suppose now instead that f is (int S)-convex; write $s > 0$ to mean $s \in \text{int } S$, and $s \geqslant 0$ for $s \in S$. Then for $0 < \lambda < 1$ and $a \neq x$,

$$\lambda f(x) + (1 - \lambda)f(a) > f(\lambda x + (1 - \lambda)a).$$

Therefore

$$f(x) - f(a) > [f(a + \lambda(x - a)) - f(a)]/\lambda$$

$$\geqslant f'(a)(x - a)$$

by the result of the previous paragraph. Hence f is (int S)-convex only if, whenever $x \neq a$,

$$f(x) - f(a) - f'(a)(x - a) \in \text{int } S.$$

The converse statement is readily proved, as in 2.4.3.

Consequently, as in 2.4.3, if $Y = \mathbb{R}$, $S = \mathbb{R}_+$, and f is twice Fréchet differentiable, then f is (int $\mathbb{R}_+$)-convex if $f''(x)$ is positive definite, for each x.

Exercise. If $f : X \to \mathbb{R}$ is convex (thus $\mathbb{R}_+$-convex), show that the directional derivative

$$f'(a, y) = \lim_{\lambda \downarrow 0} \lambda^{-1}[f(a + \lambda y) - f(a)]$$

exists, for each direction y.

Index